SIMPLE CIRCUIT BUILDING

Other Constructor's Guides

Electronic Components
Electronic Diagrams
Practical Electronic Project Building
Printed Circuit Assembly
Project Planning and Building

SIMPLE CIRCUIT BUILDING

P C GRAHAM

NEWNES TECHNICAL BOOKS

THE BUTTERWORTH GROUP

UNITED KINGDOM
Butterworth & Co (Publishers) Ltd
London: 88 Kingsway, WC2B 6AB

AUSTRALIA
Butterworths Pty Ltd
Sydney: 586 Pacific Highway, Chatswood, NSW 2067
also at Melbourne, Brisbane, Adelaide and Perth

CANADA
Butterworth & Co (Canada) Ltd
Toronto: 2265 Midland Avenue, Scarborough, Ontario M1P 4S1

NEW ZEALAND
Butterworths of New Zealand Ltd
Wellington: T & W Young Building,
77–85 Customhouse Quay, 1, CPO Box 472

SOUTH AFRICA
Butterworth & Co (South Africa) (Pty) Ltd
Durban: 152–154 Gale Street

USA
Butterworth (Publishers) Inc
Boston: 19 Cummings Park, Woburn, Mass. 01801

First published in 1976 by Newnes Technical Books
A Butterworth imprint
Reprinted 1978
© Butterworth & Co (Publishers) Ltd. 1976

All rights reserved. No part of this publication may be reproduced or transmitted in any form or by any means, including photocopying and recording, without the written permission of the copyright holder, application for which should be addressed to the publisher. Such written permission must also be obtained before any part of this publication is stored in a retrieval system of any nature.

This book is sold subject to the Standard Conditions of Sale of Net Books and may not be resold in the UK below the net price given by the publishers in their current price list.

ISBN 0 408 00230 1

Typeset by Butterworths Litho Preparation Department
Printed in England by Chapel River Press, Andover, Hants.

PREFACE

Practical projects require some degree of knowledge from most constructors to be sure of obtaining a successful working circuit. Theoretical and practical expertise will be obtained by making up circuits over a period of time. The layman, beginner or student looking for detailed information about circuit operation, coupled with step-by-step constructional guidance, might expect a book of this size to contain a limited number of specific electronic applications. Since this kind of approach is well covered by the popular monthly magazines, it is felt that the constructor could broaden his knowledge and experience by taking a course in more general practical circuit building so that he can adapt to his own needs some circuit modules as he requires them.

This volume is intended to provide a logical approach to general purpose circuits for experimenting. It is elementary but not basic and it is expected that the reader would acquire details of specific applications from other sources. The suggested circuits are easy to assemble, some ideas being given on converting a theoretical circuit into a practical layout. Detail on construction and component information are omitted since this information can be found in other volumes in the *Constructor's Guides* series.

In compiling this book it is inevitable that reference will be made to well established circuits. There are, however, special cases where a circuit has individual features that are not always adaptable for general purpose applications. The reader is recommended to refer for further information on specific topics to the application notes issued by most manufacturers of integrated circuits and transistors. Journals also provide up-dated information. For the examples given here the author wishes to acknowledge the help of SGS-Ates, Signetics Corporation and RCA, who provide the application notes on the linear integrated circuits shown.

Although this volume is not as comprehensive as might be expected, it goes some way to filling a gap in published information for constructors. If it provides food for thought or answers some questions that arise during project work, it will have fulfilled its job. Readers are recommended to have other volumes in this series to hand for further reference.

P. C. Graham

CONTENTS

1
Switching Circuits 1

2
Logic System Modules 19

3
Rationalised Layouts for Switching Circuits 35

4
Operational Amplifiers 43

5
D. C. Power Supplies 62

6
A. C. Amplifiers 74

Appendix 89

Index 111

1 Switching circuits

By convention the student is faced with two basic approaches to understanding how electronic circuits work. It can be frustrating to spend long sessions going through the rudiments of physics and basic electrical theory, but in many instances there is a need to create a working example by building simple practical projects, such as those frequently described in magazines. How can one cut a few corners and learn from practical experience some of the methods that may escape the straightforward instructional type of information? The fact that electronics theory is based on a combination of logical analysis and practical experience defies any possible suggestion that it is an exclusively mysterious science.

It is commonplace to find textbooks that provide descriptions of electron theory and circuit analysis with a certain amount of detachment from the reality of a conglomerate assembly of hardware. In these cases it is left to the student to provide the mental agility to associate one with the other. It is the intention of this book to show how one can learn from practical examples using the minimum of resources, whilst following a logical approach. The method is basically simple, the results probably obvious; however, the idea within this book is to try to tackle problems of circuit theory without consciously looking at it from a purely academic approach.

Companion volumes in the *Constructor's Guides* series give rudimentary information on the components that are frequently used. Practical help is also given elsewhere in the series on component selection and application, so it is a good idea here to take a close look at what electronic circuits can do and why. It is obviously best from the student's point of view to start with fairly basic principles and progress through to more complex applications; it is hoped that the novel approach that is adopted here will help the newcomer to learn about what electronic circuits can do.

It is useful to be armed with the necessary tools so that some of the circuits can be assembled as one goes along. The components used are fairly common and at the time of writing are easily obtained through retail electronic components stockists or by mail order. However, before going ahead with building any of the circuits the reader should satisfy himself that he can confidently expect to achieve results using the components purchased, and that no 'substandard' or 'out-of-specification' devices are used.

Simple Electronic Switch

Probably the easiest first stage is to look into the kind of simple electronic switching circuits that are developed for computers and industrial or commercial control systems. Later we shall look into the electronic circuits that are used in entertainment equipment, such as audio, radio and television. First of all, the author makes no apology for introducing an easily recognised everyday situation.

Imagine an average family car on the road—not moving but waiting to be driven. Assuming the road to be level the car will remain stationary. As soon as the driver starts the engine and engages the gears it will move at a speed determined by the accelerator setting. He can make the car move slowly or fast and in the direction he wishes to go.

In a busy city he will encounter traffic jams, road works and various other restrictions outside his control that will influence the progress the car makes. He will also pass over detectors that control traffic lights. What, you may ask, has car driving in common with electronics?

Imagine for a moment that the road is represented by a plain line, the direction of movement of the car by an arrow head, the engine or other force that makes the car move is represented by a square. Fig. 1a shows these features. Now, if the car is slowed down due to some obstruction (Fig. 1b) or by releasing the pressure on the accelerator pedal (Fig. 1c) then the car will slow down. This can be represented by a hazard on the line (Fig. 1b) or a variable control affecting car engine speed (Fig. 1c). Round the corner there could be traffic lights which will tell him if and when he can proceed (Fig. 1d). Notice that the accelerator control is indicated by an arrow through the car box, showing that variation of speed can be arranged. The lights are operated by a timed control system that supplies an instruction or command signal. The car is allowed to proceed only when the lights are in its favour. The traffic lights will show that the car is allowed to go or must stop. Suppose that these two conditions are represented by a switch as

Fig. 1. Analogue of electronic circuit

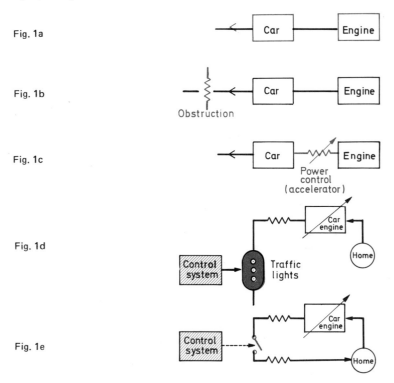

Fig. 1a

Fig. 1b

Fig. 1c

Fig. 1d

Fig. 1e

Basic analogue of the electronic switching circuit, illustrated by the motor car

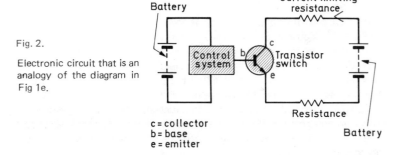

Fig. 2.

Electronic circuit that is an analogy of the diagram in Fig 1e.

c = collector
b = base
e = emitter

in Fig. 1e. If the switch is kept open, the car cannot complete the circuit back to home.

The reader may have noticed the analogy with a simple switching circuit which is shown in basic form in Fig. 2. This is the theoretical circuit diagram that shows by graphical symbols what route the electric current from the battery will take, and how it is influenced by electronic components on the way. (The companion volumes in this series on *Electronic Diagrams* and *Electronic Components* show the meaning of the symbols and show what the components can do.)

Since the car represents an electron or group of electrons, the engine represents the power to drive those electrons into motion. Just as the size and power of the car engine determines how fast the car can travel so the size and electromotive force of the battery will determine the current flow, or electron motion, through the circuit. To make anything like this work it is necessary to provide energy. In electronics the energy source can be a battery or a generator which drives the electrons into motion through all continuous paths in the circuit. A switch can be inserted in the circuit to stop current flow.

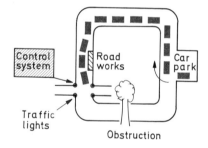

Fig. 3

Several cars on the road drawn to represent an analogy of a multi-electron circuit

Current flow is completely halted throughout the entire circuit; the switch can be positioned anywhere in the line to have the same effect.

We have seen the simple analogy of one car representing one electron but in an electrical or electronic circuit electrons throughout the circuit would be made to move simultaneously, just as if (in the analogy) there were several cars bumper to bumper. They might look like the drawing in Fig. 3. Imagine that all of the cars are leaving a car park at about the same time and following the same route.

Traffic lights tell the driver not to proceed except when green. This condition controls the flow of traffic round the whole circuit and back to the car park. As soon as the controller changes the lights, the cars can continue on their journey.

Let us now assume that the lights are green but that an obstruction has blocked the road further on. The cars can proceed up to the obstruction then stop again. If there was no obstruction the cars would be able to proceed to complete the circuit, and even continue round the circuit again.

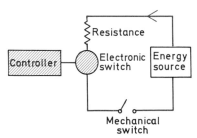

Fig. 4

Simplified form of electronic circuit analogy to Fig. 3

In electronic terms the diagram in Fig. 3 can be redrawn as in Fig. 4 to show a complete circuit with a switch and a controller. The car park in Fig. 3 is replaced by a source of energy—a battery which supplies energy or influence to create electron motion in one direction. The resistance limits the flow of electrons.

Now using more familiar terms, look at Fig. 5a, which shows the same arrangement, represented by the symbols for electronic components. The equivalent showing the actual components is shown in Fig. 5b. The controller remains as an unidentified box for the moment

Fig. 5. Basic electronic switch

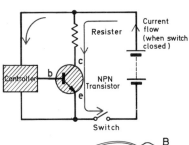

Fig. 5a. Switch using standard circuit symbols

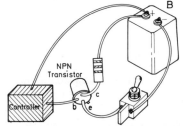

Fig. 5b. Physical arrangement of the circuit in Fig. 5a

because later we shall see how we can control the main circuit in different ways. The transistor acts as a switching device to regulate the current flow, being switched on or off by the controller. When the transistor is switched on, the current through the circuit is relatively high compared with the control current. The only limiting factor is the resistance of the collector-emitter junction within the transistor and the internal resistance of the battery, which are both very low. If the resistor were not inserted in circuit the high current would be likely to damage the transistor. All transistors are given specific current ratings which must not be exceeded.

In the circuit in Fig. 5, current cannot flow until two conditions are satisfied: (a) the switch must be closed and (b) the transistor made to conduct through the collector-emitter junction by feeding a small current into the base-emitter junction. The transistor shown is an npn type, i.e. the emitter must be connected to the most negative terminal of the d.c. supply.

Fig. 6. Current paths of transistor switches

Fig. 6a. Using an npn-type transistor

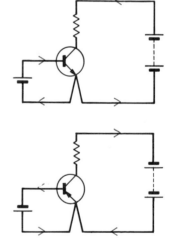

Fig. 6b. Using a pnp-type transistor

In this case the emitter is common to both the base and collector current paths and will give a clue as to how one can supply the base current from the controller. The simplest way of showing this is in Fig. 6. The base current must be of the same polarity as the collector current with respect to the emitter. If a pnp transistor is used the polarities would be changed to those shown in Fig. 6b.

Fig. 7

Equivalent circuit of that in Fig. 6a using diode symbols to represent transistor junctions

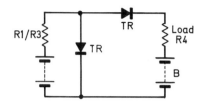

Fig. 8. Transistor switch

Fig. 8a. Using one battery supply for both current paths

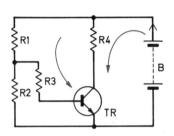

Fig. 8b. Using one battery to supply three current paths for two transistors. R1 provides the base bias current path for TR1, which in turn provides bias for TR2

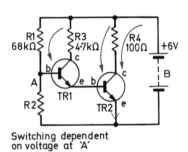

Switching dependent on voltage at 'A'

Fig. 8c. The components of the simple circuit in Fig. 8b assembled on a tag strip

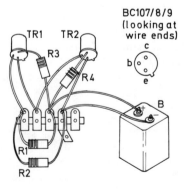

In this book we shall concentrate on circuits using npn transistors to avoid unnecessary confusion, assuming that the reader will interpret the polarities in reverse for pnp circuits. Fig. 6a may be redrawn in another way using diodes to represent the junctions of the transistor. This is shown in Fig. 7, but straight away it will be seen that there is no direct link between the two diodes to encourage one to influence the other; in a transistor that link is built in. Consequently, one cannot substitute the diodes in place of a transistor, but one can use a transistor in part to perform as a diode; this is often applied to production line assembly of some low cost consumer goods where the economical factors are in favour of doing so.

Coming back to our practical circuit, it is more usual to use a common d.c. power supply source if possible to supply all the needs of the circuit, so somehow the base and collector current paths could be derived from the same battery. Fig. 8 shows two ways of doing this; by means of a potential divider network across the supply or by extracting a current path from another transistor circuit.

In Fig. 8a, R1 and R2 form the potential divider, with a current limiter if required as R3. (In a later chapter we shall see how R3 can be replaced by another source of supply for processing a.c. signals.) The base current is derived via R1 and R3; in Fig. 8b the base supply for TR2 flows through R3 and TR1 when TR1 is in a conducting state. This latter example is very useful where a high current is required through the load resistor R4 of TR2, and only a small switching control current is available, in this case at the base of TR1. In both these examples R2 serves to provide reasonably stable d.c. operating conditions for the transistors, to offset any possible tendency of increasing current with consequent damage to the transistors. It may be recognised in various forms according to the application.

Both circuits in Fig. 8 represent simple 'grounded' or 'common-emitter' transistor stages which by themselves will probably seem to be of little or no value unless we can make them perform a suitable task. The next few chapters continue by showing the processing of direct current in simple switching circuits and gives some examples that you can try out for yourself. Later chapters show how to build up circuits that will process or amplify a.c. signals.

Practical Switching Circuits

A simple single transistor stage can be developed into a number of switching functions. The easiest way to develop an electronic switch is by arranging for a single transistor to be in a state of conduction through its collector when a battery is connected. Fig. 9a shows how this can be done, but it will not be very helpful to the constructor unless he can usefully employ the idea or demonstrate that it will work. Fig. 9b is the same circuit but to check for conductivity connect a multimeter in turn in each of the three current connection positions shown making quite sure that the correct range has been selected first.

In position M1, a very low current is expected so the current range will be set to read about 25–30 µA. The current could be higher, depending on the transistor gain characteristics, so it is better to set the meter to 50 µA, to start with, changing the range as necessary to read off the current.

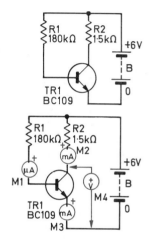

Fig. 9. Transistor switch

Fig. 9a. Basic single transistor switching stage

Fig. 9b. Connect current and voltage meters to monitor operating conditions

Make a note of this reading, then connect the meter into position M2 (Fig. 9b); R1 should be reconnected to TR1 base. The meter should now be set to a range of about 1 mA to read the collector current. If the transistor is working the current reading at M2 should be considerably higher (about 350 µA) than that in M1. The total current measured at M3 will be the sum of M1 and M2 plus a very small amount of leakage current. In all these measurements, make sure that the meter connections are correct as shown.

Now change the multimeter range to read on the voltage range up to about 10 V full scale. Restore the circuit to that shown in Fig. 9a and measure the voltage from the collector to the common emitter. For the values shown in this circuit the collector voltage (M4) will be much less than one volt and could be as low as 0.4V.

This simple experiment will tell us several things. First the current gain of the particular specimen transistor used, which could vary considerably according to the collector current. This gain is found by dividing the current at M2 by the current at M1 and will apply for the working conditions set by the values of R1 and R2. Secondly, if R1 is disconnected from the base, then no collector current will flow; thirdly the voltage shown at position M4 will be almost zero when the transistor is switched on and conducting, and at the full supply voltage when no collector or base current is flowing.

We have an effective on-off switching state through the transistor. How can this be used externally to control other circuitry? Before attempting to answer this question let us add a similar circuit to this so that we have two electronic switches. The main difference is that the additional circuit has low value resistors to enable a higher current to flow, so that some form of indicator, such as a pilot lamp, can be connected to show the collector current flow. The circuit diagram is shown in Fig. 10a. For the component values shown the collector current would be in the region of 50 to 60 mA while the base current is about 250 to 300 μA. When switched on the collector voltage would be low but not as low as demonstrated by the circuit in Fig. 9; here it will be about 1.2 V.

The significant differences in these basic characteristics show that the transistor current gain will be higher for a higher collector current. Manufacturers specify characteristics that cover a wide range of current gains found from one sample to another of the same type. However, this will be quoted at a given collector current I_c. One should always use these figures as a guide only, bearing in mind the variations that are possible for different operating conditions.

The main factor restricting operation will be the maximum current that the collecter will withstand. At best, the early symptoms will be that the case of the transistor will run warm. Unless some form of 'heat sink' is used to carry the heat away, there is a risk that the increased temperature will lower the junction resistance, so increasing the current flow. The threshold of operation at a safe temperature will depend on the size and efficiency of the heat sink as well as the operating conditions. If the designed collector current is well within the limits specified for the device used, then a heat sink is not likely to be required, especially for low current devices of less than about 100 mA. If a higher current requirement is contemplated then one would entertain the use of another transistor type able to withstand the current used.

It is possible to demonstrate collector current flow without a meter; by replacing the 100 Ω collector load resistor by a small bulb—in this case a 6 V, 0.04 A or 0.06 A bulb will suit the situation and will show when the transistor is switched on.

A bulb can be inserted in place of the 1.5 kΩ collector resistor (in Fig. 9) provided a series resistor is connected with it. However,

Fig. 10. Transistor switches

Fig. 10a. Two switches carrying different load currents through collector circuits

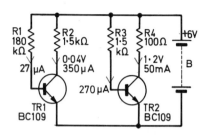

Fig. 10b. The two stages combined to form a two-stage amplified switch

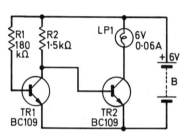

Fig. 10c. The practical circuit shown above connecting components in 'bird's-nest' fashion

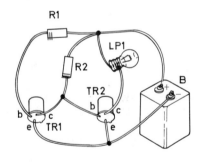

Fig. 10d. Switched input to the transistor switching circuit

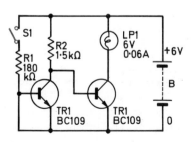

because of the low current in this circuit it is unlikely that the bulb will glow when the transistor is switched on.

The two transistor switches shown in Fig. 10a need a means of setting them into action and this is effectively done if a suitable voltage is applied to their respective base connections. With a 6 V 0.06 A (nominal rating) bulb connected as the anode load the base voltage is at 0.85 V and collector 1.45 V for a collector current of 50 mA, just about half of the maximum recommended collector current for the BC109. All that is required then is about 1 V applied to the base of this transistor.

The transistor switch on the left requires more than 0.6 V to switch it on. By linking TR1 to TR2, a small input current at 0.6 V can be made to switch on TR1 bringing down the collector voltage to less than 1 V, so holding TR2 off. As soon as the input voltage is removed (Fig. 10d) then TR1 collector will not conduct and the voltage will rise considerably. As TR1 collector is linked to TR2 base, TR2 will switch on. What is in fact required to operate this circuit is a positive current path via R1 (180 kΩ) into TR1 base. Because of the low operational current of TR1 this need not be more than about 50 μA (0.05 mA) which can easily be achieved by linking TR1 base to an automatic pulse generator or some preceding logic switching circuit. The bulb can be replaced by some other form of load such as a relay or digital indicating device, so long as the d.c. resistance is not much lower than 100 ohms.

This circuit is an electronic equivalent to an on-off switch in which a small current can be used to control a heavier load. If a changeover switch is required, whereby when one load is on another is off, and vice-versa, then two switches would be needed. In order to make them operate in alternate sequence, a pulse is supplied to each of the transistor bases alternately. While one transistor is on, it is necessary to make sure that the other transistor is kept switched off to enable a stable situation to exist, otherwise the two transistors would change over quite easily if a 'click' or other unwanted pulse reached them. To do this the transistors are cross-coupled (Fig. 11a).

Bistable switch

When the collector of one (TR1) is at high voltage (+6V in this case), a current will pass through the limiting resistor (R1) connected to it, to the base of the other transistor (TR2), holding it in a switched-on state. The only way that TR1 can now be switched on is by feeding a voltage into its base. If the base is being held by the high collector voltage of TR2, there would seem to be no way through unless either R2 is disconnected or an over-riding pulse of about 1 V is injected at TR1 base.

Set up the circuit of Fig. 11a with two bulbs and connect a 6 V battery. Straight away one of the lamps should light, telling you which transistor is conducting. Now take a 1.5 V dry cell and connect it as shown in Fig. 11b. If the left lamp is on connect 1.5 V at position B; if the right lamp is on connect 1.5 V at position A. The two transistors should change state, one switching off and the other switching on. Now alternate the connections of the 1.5 V cell from one to the other and you have a crude form of changeover switch, but of course, it would not be right to use a 1.5 V cell in this manner for practical applications because it is clumsy.

Fig. 11. Bistable switch

Fig. 11a. Switch formed by cross-coupling two transistor switches

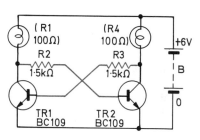

Fig. 11b. Switching is achieved by applying a positive voltage to the base of either transistor

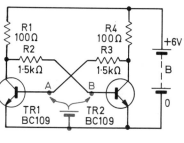

Fig. 12. Pulse switching

The bistable switch can be triggered by an external pulse via one of the two diodes

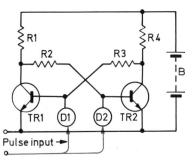

Fig. 12 shows the same kind of circuit but with an input path that will accept low voltage pulses. When the battery is connected, one lamp will light. For the purposes of this experiment, assume that the lamp (position R1) is on; then TR1 collector voltage will be low (1.5 V) and TR2 collector voltage high (6 V). The limiting resistor R2 will have the effect of presenting a low voltage to TR1 (about 0.8 V), just enough to overcome the inherent voltage drop of about 0.6 V across the base-emitter junction, so causing base current to flow in TR1.

To make TR1 switch off a pulse via diode D1 must reduce the base voltage below about 0.4 V or alternatively a pulse via D2 must raise the voltage at TR2 base to more than 1.5 V (TR1 collector voltage). The circles representing the diodes have been left deliberately undetailed because the diode polarity is an important feature in effecting correct switch-over. The connection really depends on the kind of pulse fed into the 'trigger' pulse input line and the state of the circuit polarity.

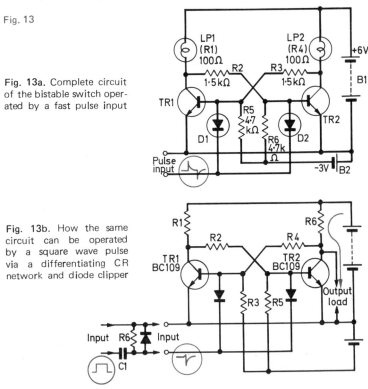

Fig. 13

Fig. 13a. Complete circuit of the bistable switch operated by a fast pulse input

Fig. 13b. How the same circuit can be operated by a square wave pulse via a differentiating CR network and diode clipper

If we assume the circuit to be as shown in Fig. 13, then it is necessary to tie down the transistor bases via current limiting resistors to a negative voltage supply, so that a negative going pulse of sufficient peak voltage can drive the non-conducting transistor into conduction, so changing the switched state of the circuit, which should lock once changed. This is called a 'bistable' switch and the loads, represented here by the lamps, are connected to the collectors.

Since only an on-off situation is needed, one can take the 'switched' line from the collector of either transistor and use the common line (from battery negative here) as the other connection. This means that the collector resistor will have to be capable of carrying the load

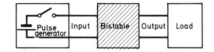

Fig. 14. Block diagram

Showing how the bistable can operate as a switch

current. Notice that the control pulse is isolated from the load by the transistor (TR1) collector which is non-conducting when the current passes through the load. When the transistor TR2 is conducting, there is an effective short circuit across the load.

This is the very simple form of switch that forms the basis of most logic switching circuits. In order to identify the switched states, it is best to use some kind of notation to describe them. The simple '0' and '1' binary numbers are used because they can, in complex as well as simple systems, give an arithmetic solution to switching problems.

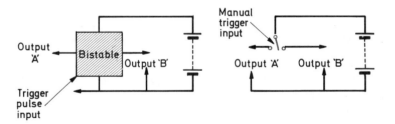

Fig. 15. The bistable can operate in the changeover mode with the positive supply to the 'wiper'

The 0 signifies that the output is off or at almost zero volts; 1 signifies that an output voltage is available to supply the load. Only a very short pulse is needed to change the switched state which will not change again until another negative going pulse is supplied at the input.

The main purpose of such an idea might remain obscure to some readers until one recollects the earlier analogy of the traffic lights. Here we have a control circuit (the bistable) operating as an electronic switch. The pulse can be supplied from a remote push button or by feeding a signal from a pulse generator (Fig. 14); a very small instantaneous voltage can be made to switch on or off a larger load.

It is possible to make use of this small pulse to control several loads, even if different switched states are required. Remember that the bistable can supply two output conditions simultaneously, so it is effectively a changeover switch, assuming the positive supply line acting as the equivalent of the wiper in a mechanical switch (Fig. 15). When output A is off (0) output B is on (1) and vice versa. So why not use a mechanical switch? The main advantages of using electronic switches are that they can be timed to a very high degree of accuracy and operated by means of unattended electronic pulse generators. Let us look at an example of how this can be done.

Astable multivibrator

Fig. 16a shows a circuit that might be easily mistaken for a bistable switch, but it is, in fact, a 'square wave' generator. This is a self-starting circuit that operates in a similar manner, but provides from its own power source an output that is alternately at maximum and zero voltage.

The changeover of the state of the transistors from one to the other is almost instantaneous and the mean output voltage the same for each positive output. The duration of positive or zero voltage can be preset to almost any time within the limitations of component values to hand and transistor operating limits. The total cycle period, during which maximum and zero voltage is provided, can also be preset, providing what is termed the pulse repetition frequency (p.r.f.) or the number of complete cycles in a given time period. A typical example of a circuit that can provide a square wave pulse output is shown in Fig. 16 and is called an astable multivibrator.

It commences operation by detecting an unbalance in the two halves of the circuit and setting itself into one of the switched states as described for the bistable, i.e. one transistor conducting and the other off. This allows one of the coupling capacitors to be charged up by the d.c. supply. It is slowly discharged via the base resistor and in so doing presents an increasing voltage across the base-emitter junction of the appropriate transistor until the threshold of switch-over occurs. By this time the collector voltage will fall and cause the other capacitor

Fig. 16. Astable multivibrator

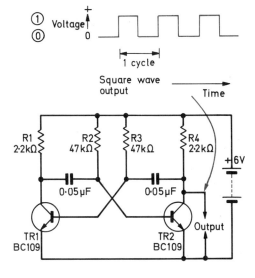

Fig. 16a. The self-starting astable multivibrator used as a square-wave generator

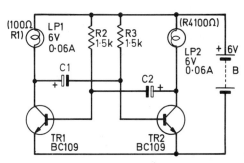

Fig. 16b. Circuit diagram of an astable multivibrator connected as an alternate beacon flasher

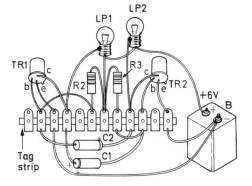

Fig. 16c. The physical wired version of the multivibrator assembled on a tag strip

to be charged. The same procedure takes place with this second transistor stage and recycling occurs through the cross-coupling of the capacitors.

This is also known as the free running multivibrator and is probably one of the most common electronic circuits used. The square wave output can be taken from the collector of either transistor, one being a positive voltage (logic 1) when the other is at near zero voltage (logic 0). The conditions are reversed when switching occurs due to charging and discharging of the cross-coupled capacitors.

This simple circuit forms the basis of a self-starting square waveform generator which is frequently used with logic switching circuits and timers. In computer and other logic systems it is also called a 'clock-pulse' generator (CP) since it is the heart of the timing section.

As a novelty device the multivibrator is often used to provide alternating flashing beacons using low voltage bulbs (Fig. 16b) or as a metronome providing beat timing through an earphone or small loudspeaker. The multivibrator is also shown as a point-to-point wiring layout for printed circuit and direct wiring comparison (Fig. 16c). Of course, component values will vary from one application to another, depending largely on the timing factors required and the load current expected. These projects are shown merely to illustrate the different principles so that they can be readily recognised.

2 Logic system modules

We have seen from the previous chapter how an electronic circuit can perform automatic switching functions; the introduction of common logic notation, i.e. binary numbers 0 and 1, is an important elementary stage in deciphering complex switching circuits without being confused over voltages and polarities. Of course, it must be established at the outset just what 0 and 1 are to represent and one must maintain consistency throughout. The most common notations are shown in Table 1.

Table 1. Examples of binary notation

Logic code	
0	1
Off	On
No	Yes
Reverse	Forward
Zero	Positive (positive logic)
Zero	Negative (negative logic)
Left	Right
Incorrect	Correct
Low	High
Down	Up
Pull	Push
Dark	Light
Black	White

Most of these situations could be answers to simple questions involving no more than two alternative possible answers. In the list there is included positive 1, zero 0, negative 1 which might seemingly add to the confusion where both positive and negative (e.g. voltages) are involved. However, it must be established at the outset if the electronic circuit uses positive or negative voltages; zero will always be made the reference, so can be 0 in all situations.

The importance of establishing a logic notation becomes obvious when we use, for example, a free-running astable multivibrator to provide positive-going pulses for a complex system.

If these pulses are to be routed through various paths to trigger or switch other electronic circuits into action, then some kind of routeing code may be necessary to ensure correct timing or sequence of operations. Maybe an inversion of switch code is required, or even the ability to supply several loads with the same pulse simultaneously. This is where gating circuits are very useful, because they can offer a number of functions at once, either on their own or with buffers.

Gates

The simple gate has to recognise the kind of pulse received and provide a similar pulse at the output at the right time. As an example, look at Fig. 17 which shows a plain box, two input paths and one output. Table 2 shows what might happen at predetermined time intervals which for simplicity are marked with numbers.

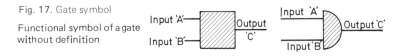

Fig. 17. Gate symbol
Functional symbol of a gate without definition

Suppose the gate was required to supply a pulse at the output when a pulse is fed into either or both of the inputs. Sequence row (a) shows what happens when one pulse is followed by another. Notice that if we use 1 to show a pulse and 0 for no pulse, the definitions are quite clear. In order to get an output pulse at a given instant, input pulses must be supplied as indicated by 1's. There will be similarities in some conditions as shown but qualification and definition is indicated by the full sequence of alternatives for a pulse at neither input, one input, and two inputs.

These are generally called 2-input gates and can be an OR gate where a pulse at either input will provide a pulse at the output; a NOR gate which is the inverse of an OR gate providing no pulse at the output when a pulse is fed to either input; an AND gate where a pulse must be applied to both inputs simultaneously to provide an output pulse; a NAND gate where a pulse at both inputs will provide no pulse at the output.

Table 2. Positive notation of a 2-input gate

	Output function		1 ABC	2 ABC	3 ABC	4 ABC
	Inputs	Output				
(a)	A OR B→	1	000	101	111	011
(b)	Neither A NOR B→	1	001	100	110	010
(c)	Both A AND B→	1	000	100	111	010
(d)	Not A AND B→	1	001	101	110	011

A simple gate circuit is shown in Fig. 18 which is the basis of gating circuits that are made under the general group called 'Diode-Transistor Logic'—or DTL—available in integrated circuit package. Since it is much cheaper and easier to use i.c.s. for this purpose there is no advantage in making gating circuits by building from a number of discrete components.

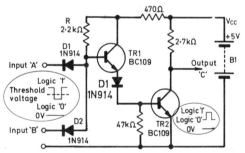

Fig. 18. DTL NAND gate

Circuit diagram of a 2-input NAND gate in the Diode-Transistor-Logic (DTL) range of i.c.s

The circuit shown in Fig. 18 is a 2-input NAND gate which is generally found in DTL i.c.s. It is possible to arrange the required function using this type on its own and with other NAND gates and inverter circuits, the latter providing inverse logic outputs. Such rationalisation for i.c. manufacture is both economic and sensible. If the constructor requires other functions than NAND he can either use these further circuits or build his own. However, for the purposes of this book it is recommended that ready made i.c.s are used, since consistent performance and near perfect circuit matching can be achieved.

The alternative gate i.c.s in the Transistor-Transistor Logic (TTL) range are also recommended. These are very similar to DTL gates except that instead of having individual diodes at the inputs, the input transistor has a built-in array of a number of emitters to do much the same job. This is shown in Fig. 19 where again NAND gates are made in the rationalised scheme.

An earlier series of integrated circuit gates were produced in the Resistor-Transistor Logic (RTL) family, in which each input pulse is fed via an input 'resistor-transistor' stage, but this system is seldom used now because of the relatively slow speed of operation, comparatively poorer output pulse wave shape and higher cost.

These gates are not limited to only two inputs; for convenience of manufacture and applications the range is made with one, two, three or four inputs, although in special applications more may be used, or a further group of diode inputs can be connected to a common expander input.

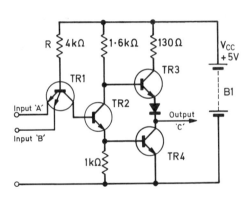

Fig. 19. TTL NAND gate

Circuit diagram of a 2-input NAND gate in the Transistor-Transistor-Logic (TTL) range

The advantages of integrated circuit gates are rarely appreciated until a full understanding of all their characteristics is achieved. There are a number of excellent books that set out to provide more of the theoretical detail including *Integrated Circuit Pocket Book* by R. G. Hibberd. Let us look here at the more practical points.

NAND gates are designed so that the input pulses have a positive-going voltage derived from a matched preceding circuit or pulse generator. It is probably easiest to see from Fig. 19 that a current path, via the 4kΩ base resistor and the two emitters of the input transistor, will be needed to make the circuit switch. This is called the 'current sinking' path and is completed by the output transistor of a preceding stage. This is shown in Fig. 20a for TTL and Fig. 20b for DTL.

These two circuits show two similar gates connected together, but so long as the correct current sinking path is provided for the input, then gate switching can be achieved. There is always a threshold input voltage level that will determine the ability of the gate to switch, based on the natural resistance of the input diode junction(s).

For most purposes it can be safely assumed that a pulse of more than 1.4 V for DTL and TTL will provide switching. Input sinking current will be about 10 mA maximum for DTL and TTL depending on the series resistor R (Figs. 18 and 19). To avoid junction breakdown, this resistor is vital and if not already provided in the i.c. package, it must be inserted in circuit.

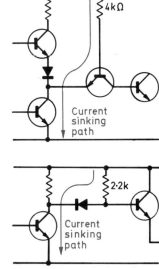

Fig. 20. Current sinking paths

Fig. 20a. TTL

Fig. 20b. DTL. Both diagrams show current sinking paths from the input of one gate through the output of the preceding circuit

With the arrangement shown an output pulse up to near maximum supply voltage can be made available with a load current rating of 30 mA maximum for DTL gates or 10 mW for TTL. These figures are often translated into the number of matched equivalent circuits that can be connected to the output, called 'fan-out', which is about ten. For higher output loading a buffer amplifier or inverter stage must be added. This is sometimes included within some i.c. packages such as bistable switches (otherwise called 'flip-flops') so that loads of up to three or five times can be accommodated. It is possible to use DTL and TTL circuits together, as their basic input and output requirements are very similar; this is known as compatibility.

Now let us look at a simple practical gate switching circuit used with the astable multivibrator and bistable switch. Constructors may like to try experimenting with the i.c. and discrete component versions. Make up the basic circuits for the astable multivibrator and bistable switch and make sure that each section functions before interconnecting the inputs and outputs.

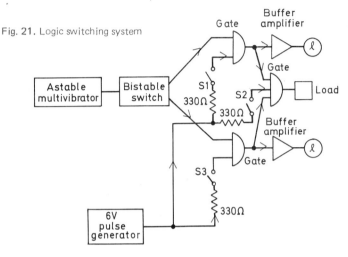

Fig. 21. Logic switching system

Block diagram of a logic switching system triggered from a multivibrator and 6 V pulse input. Logic states are indicated by lamps

By following the guidelines later in this chapter, build up the complete system as shown in Fig. 21 and incorporate a battery supply path which can also be introduced as a pulse input into one input of each of the gates, so pre-conditioning the gating. The buffer amplifiers are added so that indicator lamps will show the output state of the gates. The whole system can be driven from a 6 V lantern battery, which is also used for the pre-conditioned inputs. It is probably best to build each circuit block as an individual module, then connect the inputs and outputs together as required.

Buffer amplifier

The circuit diagram for the buffer amplifier is shown in Fig. 22. This is a simple single stage d.c. amplifier that will indicate the output logic of each gate; when the lamp lights, a '1' is indicated and when the lamp is off a '0' is indicated. It is a very useful little module to make as it can be used to provide a quick check of logic state at almost any stage in building up a complex logic switching system. Of course, it is important to observe the correct polarity of the battery supply and to

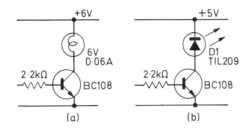

Fig. 22. Buffer amplifier

For use with (a) low current filament lamp, and (b) light-emitter diode, indicators

connect to logic circuits with positive power supplies (positive logic code). If required for negative logic circuits then the common line will be the +6V supply and the input will remain the same.

Light emitting diodes

Since light emitting diodes are fairly inexpensive and virtually everlasting, they can be used in place of the larger filament bulb with the added advantage of low power requirements; the common illuminating type such as the TIL209 is rated at 20 mA 2 V. A suggested circuit using this device is shown in Fig. 22b. L.E.D.s will only light up when correctly polarised, i.e. when connected with the cathode to the more positive supply line, so check before wiring in that it is the correct way round. This is easily done by connecting the probes of an ohmmeter across the l.e.d. It will indicate the lowest resistance when the cathode is connected to the positive probe of the meter. If the l.e.d. fails to light up in the circuit, then reverse the connections to it. These devices will light up from a 1.5 V dry cell, which will make this a very small and convenient form of polarity or logic indicator to use.

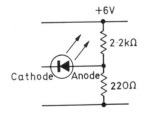

Fig. 23. L.E.D. indicator

Light-emitting diode logic indicator that operates without a transistor amplifier

A simple application of the l.e.d. is in the output of a logic circuit. It does not need an extra transistor or power supply. Fig. 23 shows the circuit of this arrangement. If the resistance network is connected across the battery supply lines, the current drain will be approximately 2.5 mA when indicating 0 (no light) and approximately 20 mA when indicating 1 (l.e.d. lit up).

This simple indicator is suitable for the switching circuits shown in this chapter and would be just as useful for DTL and TTL integrated circuit gating systems, even though these operate at a nominal d.c. supply of 5 V.

The principle is similar to the input section of a DTL gate, except that no transistors are necessary because a voltage applied to the l.e.d. will 'bias' the diode into forward or reverse current state, i.e. will pass or block the current respectively. When forward biased, the cathode terminal will be applied to a more positive voltage than the anode terminal, therefore current will flow through the l.e.d., causing it to light up. Its resistance in this forward conducting state will be low (100 Ω for the TIL209).

The more positive cathode terminal will apply when the logic state of the gate is at 1, or maximum supply voltage. When at 0 the output applied to the l.e.d. cathode will be approximately 0.7 V. The l.e.d. itself will cause a voltage drop when current flows, which is based on its forward resistance; therefore, the cathode needs to have at least 2 V applied when the minimum voltage at the anode is +0.55 V. The 2.2 kΩ resistor in the voltage divider network acts as the current sinking path for the 0 logic state, just as in a DTL gate, while the 220 Ω resistor is a current limiter for the 1 state.

Construction

Now that we have a planned system as an example to work from, let us see how this can be converted into a physical model. First, collect all the components required and lay them out in the appropriate positions on the circuit diagram; to avoid soiling the book place a sheet of transparent polythene over the page or better still trace the circuit diagram so that you can fix the tracing down on to your work bench. Place the components on this so that they are easily identified and a check can be made for omissions.

If you study the circuit diagram carefully it will become apparent that the physical wiring layout can be made very similar to it. It is common practice, but not necessarily any particular standard, to draw the circuits as shown and for constructors wishing to read more about this aspect, the companion Constructors Guide on *Electronic Diagrams* will be helpful.

There are several alternative methods, including printed circuit board, perforated board and tinned copper wire connections, Veroboard and point-to-point wiring. An example of point-to-point wiring is shown in Figs. 10 and 16 which may be quite adequate for very simple circuits. If you are contemplating making up a number of functional modules such as shown in this chapter, then it will undoubtedly be an advantage to adopt a more permanent and neat form of layout and wiring.

Since this system is not too critical in its layout requirements the choice is very wide. Let us first assume that the circuitry will be made up using discrete components (no integrated circuits). There is no reason why one should not arrange the layout as in the circuit diagram and use tinned copper wire connections or printed circuit techniques. The illustrations in Figs. 24 to 28 assume the adoption of wire interconnections and for beginners this is probably the best choice to start with. If the constructor feels inclined to be more adventurous, it is a fairly easy matter to convert the wiring layout directly into a printed circuit pattern, bearing in mind the advice given in other Constructors Guides.

There is no advantage in persuading the constructor that any one method is going to be better for him than another; what is more important is to obtain a sound working circuit at the finish. By experimenting and experience in the different methods, the constructor will always find the most suitable for his purposes. There will be other factors to consider in deciding on a suitable layout technique, such as cost, size, risks of interaction, etc., some of which are mentioned later. For example, circuits operating at very high frequencies will require special attention to component positions. Switching circuits at relatively low speeds are far less critical, but as switching speeds increase then very short interconnections become more important.

Point-to-point wiring of two-transistor circuits in the multivibrator family should be considered as about the most complex from the constructor's point of view; anything more complex should really be designed for a prepared layout panel, which could incorporate a logical positioning of input and output terminals and power supplies. By placing all inputs at the left and all outputs at the right-hand end of each panel it is much easier to find these and to follow through the sequence of events of a system employing several circuit functions.

Fig. 24 shows the astable multivibrator circuit with the interconnections between components shown in black only. This particular kind of circuit can be laid out as in the circuit diagram provided allowance is made for insulating crossed wires. Here the crossed coupling connections cannot be faithfully reproduced on a printed circuit pattern unless one is sleeved wire or a conductor passed over the reverse side of the board. Fig. 24 shows an apparent 'break' in one of the cross-coupled connections. The small circles indicate connection points, so a link wire can be connected on the component side of the board. This and the component symbols are shown in colour.

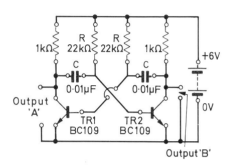

Fig. 24. Multivibrator layout

Astable multivibrator in copper wire layout form. The connections are shown by black lines. Similar printed circuit layouts can be devised

Now let us do the same for the bistable switch shown in Fig. 25. Notice here that advantage is taken of component positions in order to overcome crossover routes; the resistors to the negative supply line show an example of this in this circuit.

Added refinements to the circuit are often found in practice where 'speed-up' capacitors are connected in parallel with the cross-coupled resistors, and where a differentiation network is used to convert the square wave input into a fast pulse. These features are included in the refined layout in Fig. 26.

Monostable multivibrator

We shall see how we can take advantage of similar layouts in due course by rationalising to a common layout pattern, but first let us not forget that there is a third member of this family of cross-coupled two transistor switching circuits—the monostable multivibrator. This circuit, shown in Fig. 27, looks like a cross between the other two previously described. The basic difference is that it will change state when a pulse is fed into the input and charge the coupling capacitor. For a momentary pulse, the time that the circuit remains in a changed state will depend on the rate of discharge of the capacitor. When fully discharged, the circuit reverts to its normal inactive state. The output at the collector of TR2 will be a d.c. positive square wave pulse of duration equivalent to the discharge rate of the capacitor.

Fig. 25. Bistable layout

Using the same style of wiring layout as in Fig. 24 to make a bistable switch

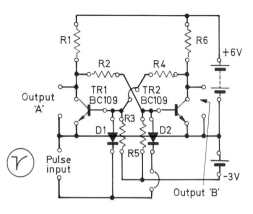

Fig. 26. Fast bistable

The bistable circuit with additional coupling capacitors in parallel with the coupling resistors to speed up the switchover. The input pulse is square shaped fed via a differentiator

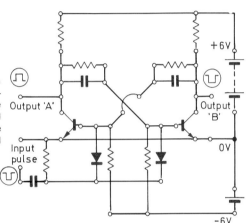

Fig. 27. Monostable layout

The monostable multivibrator uses one coupling capacitor and one coupling resistor. The initial switchover period is timed by the capacitor, then restored to the initial state after complete discharge

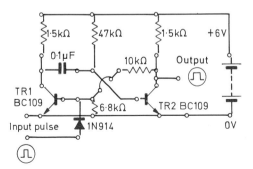

29

Fig. 27 shows how this circuit can be converted to an equivalent wiring layout, adopting similar principles as before. It is now becoming obvious that these three circuits have several common features in layout design. It becomes a relatively easy matter to make a common wiring pattern or printed circuit design that can be used for all three, and this aspect is shown in more detail later.

The student should be sufficiently confident now to choose his own wiring method for the exercise in hand and, by following the same layout ideas in principle, he can accomplish the individual construction of simple circuit modules. These methods can be used on printed circuit layouts, perforated board and copper wire, or with some slight adaptation as necessary on proprietary assembly blocks such as 'T-Dec'. Adapting to copper strip layouts, such as Veroboard, is also possible but not as easy because the copper strips may need to be cut or linked to achieve the correct interconnections.

Fig. 28. Three input logic gate

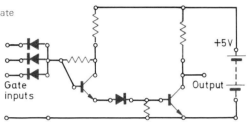

Three-input logic gate layout using two transistors and triggered by the input pulses applied via the input diodes. More details are given later

Gate inputs

+5V

Output

The gating circuit is shown in principle as a basic layout diagram in Fig. 28. This method is applied to DTL discrete circuits only because TTL discrete circuits are not viable unless multi-emitter transistors are available. It is likely that constructors will choose to use integrated circuits for these applications. Where more complex versions of any of these circuits are required then the additional layout wiring can be made on the same board.

Improving multivibrator action

The circuits shown for pulse switching are the more common arrangements likely to be found in practical applications. However, where the performance is less than adequate for a particular application, it may be worthwhile experimenting, for example, to improve the speed of switching or the square wave shape, or perhaps to provide for variability of frequency or timing. Modified versions of these circuits are certainly worth trying and are shown in the following pages.

Whereas the cross-coupled multivibrator often provides sufficiently adequate square waveform output for simple test purposes, it can suffer from having a slow switch-on time as shown in Fig. 29a by the leading edge. This is exaggerated to illustrate the point, the fact being that the curved shape of the leading edge is not well defined for determining switch-on time. Fig. 29b shows a steep leading edge having a well-defined switching time. This point is well shown up on an oscilloscope display; in the very best case the leading edge would be so fast (steep) that it would be almost invisible on the display.

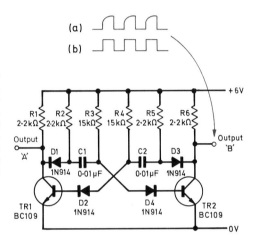

Fig. 29. Improved astable

An improved version of the astable multivibrator to speed up the switchover by means of the extra diodes. The waveforms of the normal (a) and speeded-up (b) versions are shown above. A rise time of less than 1μs can be expected

The circuit in Fig. 29 provides a very fast leading edge by virtue of the extra diodes in the coupling. A rise time of less than one microsecond can be achieved with a switch-off (fall time) of about 20 nanoseconds. In this, and the earlier simple circuit, the frequency of the square wave is found from the expression $f = 1/(1.38CR)$ where C is the value of one coupling capacitor in farads and R is the value of the base-to-supply line resistor, 15 kΩ in the example. The output can be taken from either collector.

The circuit can be used for variable frequency operation but it is better to use the emitter-coupled multivibrator for the reason that in the cross-coupled type the capacitor value should be the variable feature, which is often impossible to achieve except by using preset switched selection. The base resistors are sometimes made up into a potentiometer arrangement, but at the risk of taking the operating current of the transistor dangerously outside its recommended range.

The emitter-coupled multivibrator (Fig. 30a) can provide variable frequency by using only one variable capacitor at C without risk of saturation of the transistors. Consequently higher operating frequencies are possible. There is one snag to this circuit which is highlighted when used with other simple single polarity supply modules. The emitter-coupled multivibrator requires both positive and negative supply lines, so the output voltage may vary. The output waveform will resemble that in Fig. 30a and the frequency is not dependent on the load at the output terminals.

Fig. 30. Emitter-coupled astables

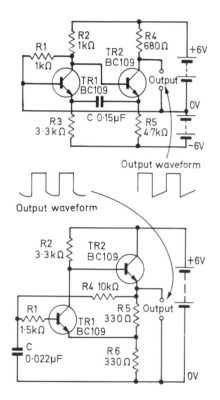

Fig. 30a. By using an emitter-coupled multi-vibrator only one capacitor is needed, so making fixed frequency of operation easier to determine. Higher frequency operation is possible because the transistors do not run into saturation. Note, however, the need for both positive and negative supply lines

Fig. 30b. Another single capacitor emitter-coupled multivibrator that uses only the positive supply. Rise time of this circuit is about 200 ns

These modified versions of the astable multivibrator will not fit into the pattern of component layout and wiring that has already been discussed, so individual design layouts will be needed. However, there is no reason why one should not follow the general idea of laying out according to the circuit diagram. Adjustment must be made where the circuit detail differs, paying special attention to the supply polarities.

Fig. 31. Emitter-coupled and complementary monostables

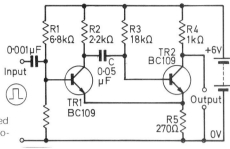

Fig. 31a. The emitter-coupled principle applied to the monostable circuit

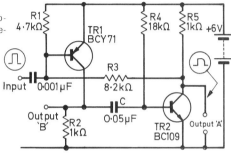

Fig. 31b. Complementary monostable circuit provides fast recovery time

Fig. 32. Complementary bistable and u.j.t. multivibrator

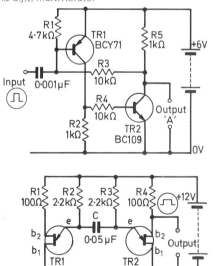

Fig. 32a. Complementary bistable switch, very useful where only one supply polarity is available

Fig. 32b. A single capacitor multivibrator using two unijunction transistors. The near square wave output at base 2 of TR2 alternates between about 11 and 11.5 V

33

The emitter-coupled arrangement can also be used for the monostable multivibrator (Fig. 31a). The advantages are the same as in the astable multivibrator although the circuit is distinctly different. It can be laid out on the same kind of wiring pattern shown for the multivibrator family, shown earlier.

The conventional monostable multivibrator shown previously suffers from a long recovery time between pulses due to slow capacitor discharge effects. To overcome this a complementary circuit using a pnp and an npn transistor can be used. This circuit, shown in Fig. 31b, differs from the previous examples in that both transistors are on or off simultaneously, conducting when the output at A is at low voltage. There is also an output point at B which will be nearly the full supply voltage relative to the common line at the same time. A positive going input pulse will switch off TRI, the collector voltage falling and turning off TR2 via the timing capacitor C. TR1 is held off until C has charged to such a voltage as to switch on TR2 again and hence TR1 also.

A similar complementary circuit can also be used for the bistable switch; an example of this is shown in Fig. 32a. A simple multivibrator using two unijunction transistors is shown in Fig. 32b. This uses the special 'negative resistance' characteristic of the u.j.t. to switch from the 'off' to 'on' state.

The timing is determined by C and the output is in the form of two 'close' voltage steps, rather than being completely off or on. It is therefore useful as a 'staircase' waveform generator or ring counter.

3 Rationalised layouts for switching circuits

There are cases where a constructor may choose to build a project that uses one or more of the more common circuit blocks. Of course, component values will not always be the same for every application, but a great deal can be done to make layout design easier if it is known that certain circuit blocks are likely to be used in several applications. It is very easy to make a rationalised layout that will cover many different applications. The advantages are that it becomes reasonably economical to produce a number of printed circuit boards with a common pattern that can be easily used for several applications. Furthermore, where a repeat circuit is required in one project, the same pattern can be ready to use and just adapted for individual circuit component values.

We have seen in the last chapter how basic similarities occur in different switching circuits. In this chapter we shall see how to derive a common pattern for many circuits so that printed circuit production can be contemplated. Precise layout and size of individual modules will probably vary according to the physical aspects of space available. However, it is intended to show what can be done by using a ready made layout design scheme. The examples given are not intended as being the ultimate for direct copying; indeed Copyright Acts prohibit such use of published material. If, however, the constructor wishes to adopt any of the designs for his personal and private use, for example, as an academic exercise or for private project work, then there is no fundamental reason why he should not do so. But as in other aspects of Copyright law, imitation or copying for commerical gain is not permitted.

First take a single transistor switch as described earlier; the layout could be something like that shown in Fig. 33 where the connections are shown in black. Note that a positive supply line and common or 'earth' line will be required in all cases. Terminals for battery or other d.c. power source connection are shown on the right as it is always best to keep power supply lines away from signal input terminals.

Fig. 33. Wiring layouts

Fig. 33a. Single transistor stage arrangement showing the wiring layout in black, assuming emitter connection to the common supply line

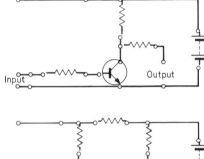

Fig. 33b. Two transistor stage layouts. The resistor values and connections can be varied as required

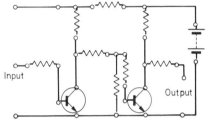

If using a perforated board, or if designing for a drilled printed circuit board, the holes will be represented by the small open circles on the diagram. Ideally, it is worth bearing in mind the likely hole arrangement and spacing from the start, since it will be much easier at a later date to fit components with fixings or connections based on a standard spacing. The most useful and common hole spacing is on a grid of 0.1 in (2.5 mm) pitch, and multiples of this. Fig. 33b shows two transistor stages. In all these examples it is assumed that 'grounded-emitter' operation will apply. Where coloured lines are shown there is always the possibility that these will represent plain wire links for some circuits, so it is always worth maintaining the facility, even if not required immediately. It is easier to remove and replace a wire link than to try cutting and peeling a printed copper foil.

From Fig. 34 the general pattern should be apparent. There may well be the need at some time to allow for an emitter resistor, such as in the Schmitt trigger circuit so this should be included. There may be a need for a feedback path although this is less likely in switching circuits except in the special case of the multivibrator family. Power line terminals at the left-hand end are very useful for linking to other similar modules. There may appear to be more connection holes than perhaps is thought necessary, but this is useful in many circuits.

This basic plan will suit many simple circuits with one or two transistors. It is probably best to work in blocks of two-transistor layouts for optimum versatility. This design has no facilities for overlapping or cross-coupled interconnections. However, if we want to adapt it for use as a gating circuit, simply link the diode input array to the pattern. If this is kept electrically separate from the remainder of the circuit pattern, it can be usefully employed for gating or a capacitor-resistor network or other device for any part of the circuit as required. The layout is shown in Fig. 34a, which is a Schmitt trigger. For connecting both emitters directly to the common line, use link wires.

Now let us consider a modified version of this layout. Fig. 35 shows the same layout but with the section around the first transistor base reversed left to right. This is the next step towards catering for the multivibrator family. However, it does not quite provide adequate connection points for these. Fig. 36 shows a more versatile multivibrator layout, which can be used for the astable, monostable and bistable circuits. The simple switching circuits shown previously can also be accommodated on this layout. Notice how crossovers are easily catered for by running conductors between component mounting holes. A printed circuit style of layout can be drawn up from this pattern, bearing in mind the usual requirements to be met regarding track width and spacing (see the Constructor's Guide on *Printed Circuit Assembly*).

The input network pattern shown in Fig. 34a can also be applied and would, in fact, be useful for accommodating the differentiator network or gating diodes. The extra supply line at the bottom of the layout is provided to carry the negative supply for the monostable and bistable multivibrators, assuming that the normal battery supply to the transistor collector is at a positive potential.

Integrated circuit packages offer even greater scope for rationalised layouts and will be particularly useful where several gating packages are used. There are several commercially made printed circuit boards available that will accept a number of 'dual-in-line' integrated circuit packages. These are straightforward and without 'frills', being intended for almost any of these i.c.s; additional interconnection wires must be soldered to the copper connection pads. These are very useful up to a point, but they have certain limitations where a custom design layout could provide a neater assembly without the familiar jumble of connecting wires.

Fig. 34. Versatile two-transistor layout

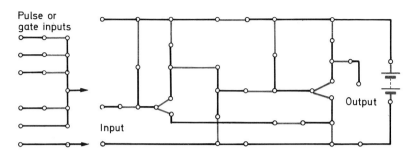

Fig. 34a. Two-transistor layout incorporating positions for emitter resistors. This circuit shows possible component positions as coloured lines. The layout is suitable for gates, Schmitt trigger, emitter-coupled multivibrators and amplifiers. On the left is shown an input network for gates, filters, clippers and clamp circuits as shown below

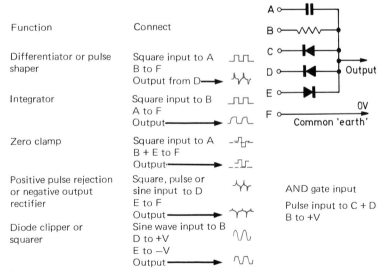

Function	Connect	
Differentiator or pulse shaper	Square input to A B to F Output from D →	
Integrator	Square input to B A to F Output →	
Zero clamp	Square input to A B + E to F Output →	
Positive pulse rejection or negative output rectifier	Square, pulse or sine input to D E to F Output →	AND gate input Pulse input to C + D B to +V
Diode clipper or squarer	Sine wave input to B D to +V E to −V Output →	

Fig. 34b. Alternative input components to provide a selection of pulse processing for almost any kind of logic circuit module

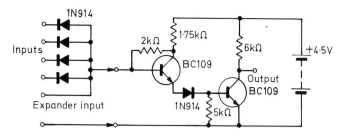

Fig. 34c. Circuit diagram of a DTL NAND gate with four inputs and an expander input for further diodes. Discrete components can be used as shown

Fig. 35. Multivibrator layout

The circuit of Fig. 34a, but with the first transistor (left) transposed to form a layout similar to the multivibrator circuit

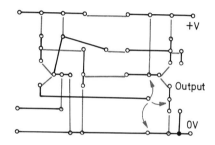

Fig. 36. Layout pattern

Versatile layout upon which several two-transistor circuits can be built. This pattern can easily be adapted for printed circuit layouts. A suggested bistable circuit is shown by coloured lines representing components. Notice how components and link wires bridge permanent interconnection strips

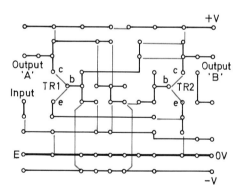

When designing complex systems to incorporate several circuit modules on one board, the designer will lay out the circuitry according to the individual components used. However, it is possible to make up 'standardised' module layouts that only need to be transferred to the required board. The layouts shown for discrete circuits can be adopted or i.c. layouts can be used.

Fig. 37 shows how a DTL i.c. containing two 4-input gates can be given a modular connection layout. How can this be multiplied or mixed effectively with layout designs for other circuits using discrete components? Notice first of all that maximum advantage is taken of the possibility of running power supply lines underneath the i.c. package and between the two pairs of pins. They are kept out of the way of signal connections.

Fig. 37. Dual 4-input gate package

Dual 4-input gate package looking at the top. All interconnections are on the underside of the board

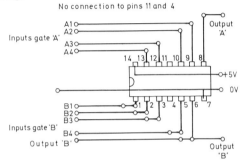

Fig. 38 shows how this method might be adapted for several i.c. packages of the dual-in-line type, each having two gates. For other types of logic gate i.c.s the connection arrangement can be modified to suit. Figs. 39 and 40 show other examples of gate i.c.s.

Fig. 38. I. C. mounting board

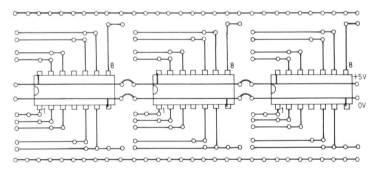

Three gate packages arranged in sequence so that interconnection is made easy. Extra components can be added if required and additional 'bus-bar' rails are provided for power supply connections

Most logic systems combine the use of different types of circuit, including some of the more specialised integrated circuits such as binary decoders and frequency dividers. The pin connection arrangements for these vary so much from one type to another that rationalised layouts become more difficult to achieve. It is probably better in these circumstances to use the ready made layout boards that have a simple series of copper pads for each pin of a number of dual-in-line i.c.s. It is common to find these arranged in parallel rows, 'military' style, each

Fig. 39

Layout for DTL series quad 2-input gates

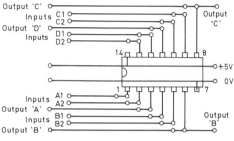

Fig. 40

Layout for DTL buffers or inverters

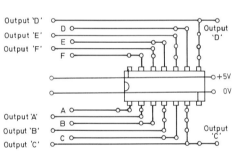

Fig. 41. Printed circuit pattern

For one dual in-line i.c. package; hole pitch 0.1 in (2.5 mm) recommended to allow four holes per pad minimum. The supply strips pass under the package

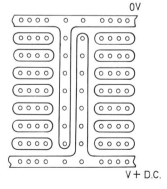

41

pad having three or more holes for connections. Three is the absolute minimum number of holes recommended—four would be much more versatile.

An example of this kind of layout is shown in Fig. 41; the pattern would be repeated for a series of i.c.s with the supply line copper conductor strips continued for all i.c.s. Generally, these conductors are brought to one edge of the board so that plug-in edge connectors can be used.

Although this kind of layout is easy to use, it is not easy to follow the circuitry from the layout wiring. It is not, of course, suitable for other i.c. packages such as the TO-5 'top-hat' can, but most logic circuitry is based on the dual-in-line package system because of its easy practical aspects and the lower cost. One serious hazard is found when the constructor needs to remove a DIL i.c. from a printed circuit kind of layout. The pins have to be desoldered simultaneously to effect removal; for this purpose special large soldering iron heads are used to cover all pins and a suction device is used to draw the solder away from the copper. More details of printed circuit methods and soldering will be found in companion volumes in the Constructor's Guide series.

4 Operational amplifiers

There are few electronic circuits that lend themselves so admirably to rationalised layouts as the switching circuits previously described. However, those systems that comprise a number of circuit modules or 'building blocks' can be made up in much the same kind of logical manner. It is not a sensible plan to mix the components of one functional circuit section with those of a different one.

When building linear circuits, i.e. those that involve the processing or generation of a.c. signals, special problems arise that are frequently characteristic of individual circuit design. Interaction is more critical in these circuits due to capacitive or inductive pick-up. It is not always a feasible proposition to lay out a linear circuit exactly as seen in the circuit diagram. Neither is it likely that interconnection of one functional circuit to another will be as easy or even possible without severe mismatch problems.

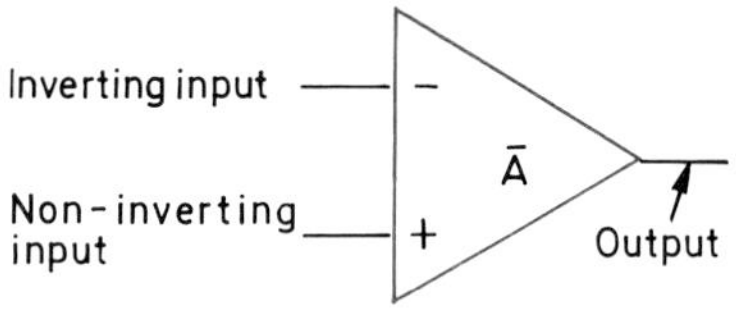

Fig. 42. Operational amplifier circuit symbol

There are familiar configurations that are easily recognised and some of these are shown in the following pages. Fig. 42 shows the functional symbol of the operational amplifier, the heart of many d.c. 'analogue' converters. The input signal is a measure of d.c. voltage or current that will provide an output related to it, that can be measured in some way or give some other kind of indication, such as operating a lamp or relay. This is different from the straight on-off state in logic switching circuits. Typical examples are as a threshold switch to operate a relay when temperature, liquid level, voltage, or strain reaches a particular magnitude.

The operational amplifier compares the input voltage with a predetermined voltage level and at the threshold causes a current to flow through the load. It can be considered as an electronic extension of the Wheatstone bridge principle that is shown in Fig. 43. When the

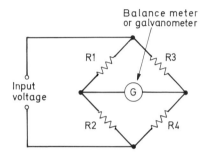

Fig. 43. Wheatstone bridge

Basic circuit to detect imbalance

resistance values are balanced, i.e. when the ratio of (R1/R2) = (R3/R4), then no current will flow through the meter load. If one of the resistance values is altered then current will flow through the meter.

In the operational amplifier, basic balance is achieved when two transistors are working in exactly the same way with identical current flow (Fig. 44). Here the transistor bases are shown disconnected and the two resistance ratios are matched. To make a usable bridge, one base must be connected to a known biasing current source and the other to a transducer, or other device, that will detect changing conditions and cause a varying current to flow.

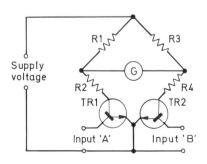

Fig. 44. Wheatstone bridge

Two transistors used to influence the balance condition of the bridge. Input voltages at A and B can be monitored and compared, the differential being indicated by the meter

Fig. 45 shows the same idea but this time TR2 base is tied permanently to a known voltage source and the input is fed into TR1 base. In order to set ideal balance conditions, if the two transistor arms are not perfectly matched, a potentiometer is inserted in the positive supply line. To provide added stability the tied emitters are connected to the collector of a voltage regulator circuit and then to the negative supply line. Even better matching stability is achieved by using dual matched low-noise transistors in one encapsulation. By keeping the operating current as low as possible the input impedance (loading of the input signal) will be kept to a level least likely to cause further unreliable operation. For the circuit shown the input impedance should be in excess of 100kΩ.

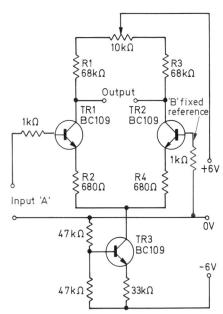

Fig. 45. Differential amplifier

Experimental differential amplifier similar to those used as operational amplifiers. TR1 and TR2 are the balance detectors and TR3 acts as a stabiliser

This comparator, or electronic measuring bridge circuit, has the added advantage of providing built-in voltage gain of about 50. It is suitable, therefore, in converting direct from a transducer to a relay, meter or perhaps a small motorised device, provided the load current is kept within the limits of the transistors' capability. If extra gain is required, or the input impedance needs to be made even higher, an extra transistor should be added to each half as shown in Fig. 46. This is generally known as the Darlington pair arrangement.

These examples are shown to illustrate the basic principle of this kind of circuit which is called by a variety of names according to its application. Constructors will find them referred to as comparators,

Fig. 46. Differential amplifier using Darlington pairs

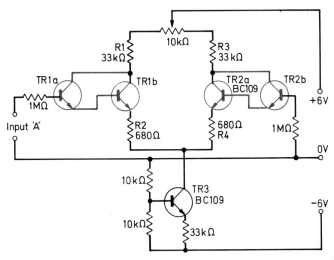

Modified version of the differential amplifier using two Darlington pairs to increase gain in the 'offset' (unbalanced) condition

differential amplifiers, long-tailed pairs, or operational amplifiers. It is seldom likely, or indeed worthwhile, to make these circuits up using discrete components as an integrated circuit package is cheaper, simpler and more reliable to use.

741 I.C. Pack

Among the many integrated circuit packages that have been developed over the years, one of the most useful is the type with the common series number 741 appearing somewhere in the manufacturers' type number. A dual version inside one package is also available under the type code 747. The circuit examples in Figs. 49 to 63 show what can be done with this extremely versatile circuit in various applications. The layouts for building them up become very simple and require little explanation, but again the constructor or experimenter will find it useful to make up a common layout that will suit several basic purposes and allow for additional components as required for the project in hand.

The integrated circuit 'op-amp', as it is popularly called, has extra features adding to its versatility, including a feedback network from the output to the input. This makes it useful for performing mathematical operations, as a filter, or as the heart of a waveform generator.

Linear circuits that perform computer functions are largely dominated somewhere in the system by the operational amplifier. This circuit usually has two inputs: one for inverting the applied signal and the other for providing a non-inverting or in-phase signal at the output. It is left to the circuit applications designer or constructor to apply extra components between the output and inputs to make it do what is required.

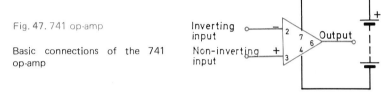

Fig. 47. 741 op-amp

Basic connections of the 741 op-amp

In the integrated circuit versions two differential amplifiers are employed to achieve a high voltage gain up to about 40 000. When negative feedback is applied with appropriate passive component values this is considerably reduced without adversely loading the output.

The circuit symbol (Fig. 47) shows the inverting input with a minus sign and the non-inverting input with a plus sign; the output is taken

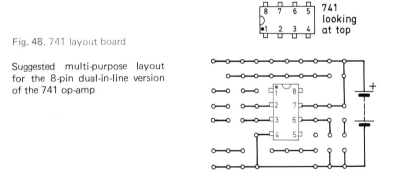

Fig. 48. 741 layout board

Suggested multi-purpose layout for the 8-pin dual-in-line version of the 741 op-amp

from the point of the triangle on the right-hand side. A suggested layout is shown in Fig. 48. With the arrangement shown in Fig. 49a, the input signal is applied to the non-inverting input although the feedback path is always connected to the *inverting* input. The gain will be equal to the ratio of the feedback resistance RF to the input resistance R2. The input signal will be a d.c. voltage or r.m.s. a.c. voltage and in this example the closed loop gain (feedback applied) will be about 20. In Fig. 49b, the input is applied to the inverting input and the closed-loop gain will be $R_F/R_1 + R_s$) or R_F/R_2.

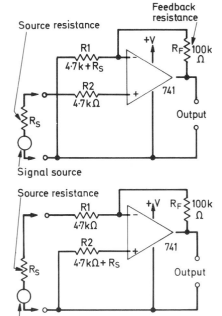

Fig. 49. Op-amp circuits

Fig. 49a. Non-inverting (in-phase) amplifier

Fig. 49b. Inverting amplifier

This is the basis on which operational amplifiers function, so it is possible to substitute any of the resistors of either circuit by reactive components, such as capacitors, so as to achieve frequency selective amplification and so convert sine or square wave inputs into a fast pulse output signal. The following examples illustrate these variations, which the constructor can try for himself using the common layout and wiring pattern already shown. As a 'breadboard' exercise, there will be many possible experiments that can be carried out; it is left to the constructor's careful selection of component values to show what results are obtained.

Following the basic examples, try out some of the applications circuits shown in Figs. 49 to 63. The true versatility of the operational amplifier will then be readily understood so that more ambitious projects can be tackled. For clarity, the power supply lines have been omitted from Figs. 49 to 63.

Fig. 50. Unity gain buffer amplifier

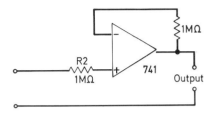

Fig. 51

Fig. 51a. Integrator

$$V_o = \frac{R_F \times V_i}{R_1}$$

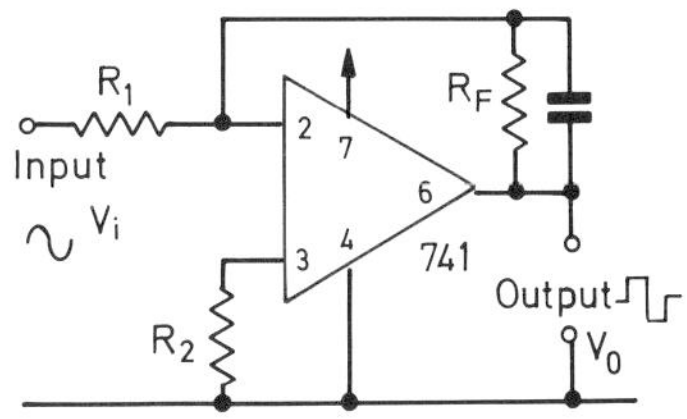

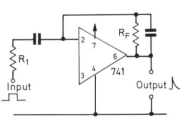

Fig. 51b. Differentiator

Fig. 52. Schmitt trigger

Threshold voltage $\simeq \dfrac{-V_o R_2}{R_2 + R_F}$

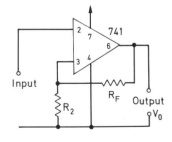

Fig. 53. Multivibrators

Fig. 53a. Astable multivibrator

$$f = 1/2C_1 R_A \log_e\left(\frac{2R_2}{R_B} + 1\right)$$

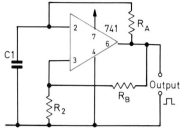

Fig. 53b. Monostable multivibrator

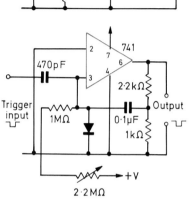

Fig. 54. Comparator

With positive feedback

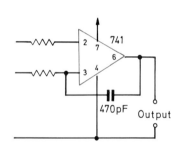

Fig. 55. Staircase waveform generator

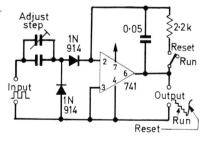

Fig. 56. Emitter follower current amplifier

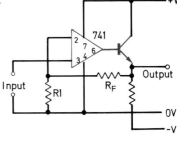

Fig. 57. Current amplifier

A development of the previous circuit using complementary output power transistors

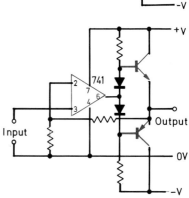

Fig. 58. Voltage adder and subtractor

Output voltage = $(V_3 + V_4) - (V_1 + V_2)$

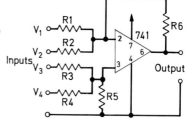

Fig. 59. Notch filter

$f = 1/2\pi R_1 C_1$
$R_1 = R_2 \cong 2R_3$
$C_1 = C_2 \cong 2C_3$

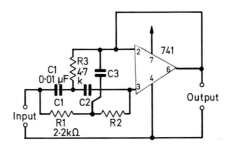

Fig. 60. Filters

Fig. 60a. Band pass filter using a 747 i.c. in a DIL package

$f_1 = 1/2 \pi R_1 C_1$

$f_2 = 1/4 \pi R_1 C_1$

$R_1 = R_2 \cong 2R_3$

$\dfrac{f_1}{f_2} : \dfrac{2C_3}{C_1}$

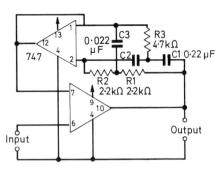

Fig. 60b. High-pass (bass cut) filter

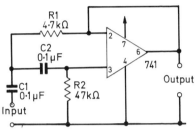

Fig. 60c. Low-pass (treble cut) filter

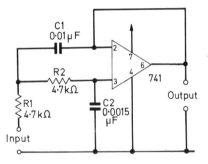

Fig. 61. Tone controls

Variable bass and treble

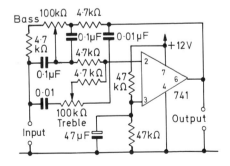

Fig. 62. Four channel mixer

With approximately unity gain up to 20 dB

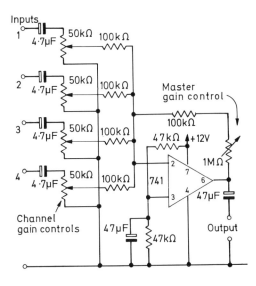

Fig. 63. Connecting the op-amp to a BC109 transistor

Fig. 63a. To an emitter follower preamplifier stage

Fig. 63b. To a current amplifier preamplifier stage

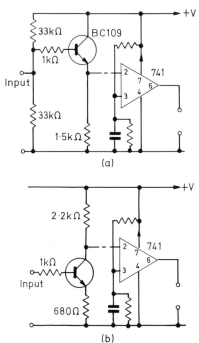

53

Do not be deterred by apparent internal complexity because you only need to know what the whole package can do to be able to make up otherwise simple projects. Many of the specification details that are published will help those using op-amps in critical designs, but for the experimenter, the basic detail on positive and negative power supply, input offset (or threshold) voltage for zero output, input impedance, output impedance, open-circuit (no feedback) gain, the effective useful frequency bandwidth (according to application), and output voltage and current are the most useful.

Practical circuits are published from time to time in various magazines which include the use of the integrated circuit version of the op-amp; the examples in Figs. 50 to 63 are basic circuits only that will help constructors to recognise these applications. A suggested layout design is shown in Fig. 48. Parallel arrangement of connection strips provides a facility for almost any application of the op-amp i.c. The circuits are available in 8-pin or 14-pin dual-in-line and round TO-5 or TO-99 packages. The input terminals are marked minus for inverting the signal and plus for non-inverting and must not be confused with the power supply lines which are marked: $+V_{cc}$, $+V$, $0V$, or $-V$.

Integrated Timer

Of more recent development is the integrated circuit multi-purpose timer circuit. The most well known type carries a type number '555' in its designation and is popularly known as the 'five-five-five' although it is generally available as an NE555 or similar. A dual version (two in one i.c. package) is designated 556. The 555 contains two comparator circuits and a bistable switch (or flip-flop) with a buffer amplifier stage at the output (Fig. 64).

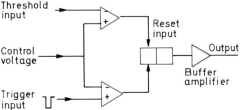

Fig. 64. 555 timer

Block diagram of the 555 timer i.c.

Like many complex or advanced technology i.c.s, this device uses basic circuit modules to provide stable and reliable performance for specific applications. Although called a timer it is useful in many cases where automatic switching is required as a result of some sort of threshold detection. It combines analogue and switching circuits and could more usefully be employed as an analogue to digital converter, in which a threshold voltage detector can be linked to strictly digital or logic switching circuitry. It is extremely versatile, condenses layout space requirements, and reduces the number of components needed as well as cost. It would be possible to make a similar circuit using discrete components or three other i.c.s, but due to matching problems the result would be a clumsy layout.

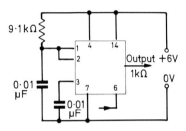

Fig. 65. 555 connections and waveforms

Typical external component connections for the 555 timer i.c. and appropriate waveforms. Pin connections shown are for half the dual version in the 14-pin package 556

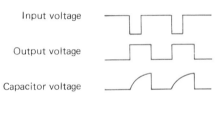

Top view of 14-pin DIL i.c. package with pin numbers

Applications of this i.c. include temperature and liquid level detection and automatic control of these, time delay switching, pulse generators and detectors, frequency division and pulse-width modulation. Time delays from one microsecond to several minutes are possible which had previously only been obtainable with complex circuitry.

To operate the timer circuit, a supply voltage between 5 and 15 V is required. The timing is set by the addition of a capacitor and resistor, the values of which will determine the timing period or time constant to allow the flip-flop to switch and allow a signal through. The flip-flop is set ready for action by the application of a trigger pulse via a second comparator amplifier, driving the output to the 'high' or '1' logic state. As the voltage across the external capacitor is allowed to increase, the first comparator resets the flip-flop and the output changes over to the 'low' or '0' state. The trigger pulse is negative-going at one-third of the supply voltage so correct supply voltage selection is probably important to the particular application.

Examples of typical applications suggested by the NE555 manufacturers (Signetics International Corporation) are given in the following pages and it will soon become apparent that circuit layouts are very simple and can be rationalised for experimental purposes. In some cases the dual version (556) will be most useful and, where appropriate, pin number connections are shown for this in a 14-pin dual-in-line package. The two separate timers are defined as A and B with pin connections as follows:

Timer A	Timer B
1. Discharge	8. Trigger
2. Threshold	9. Output
3. Control voltage	10. Reset
4. Reset	11. Control voltage
5. Output	12. Threshold
6. Trigger	13. Discharge
7. Common or ground	14. $+V_{cc}$ supply line

These connections very conveniently separate the two timer circuits so that layout wiring is easily accomplished, even when interconnection between the two is necessary.

The two comparator amplifiers can be made to produce a square wave output by connecting up one of the timer circuits as in Fig. 66 to form a free-running multivibrator. The threshold and trigger inputs are combined in the CR network comprising R1, R2 and C1. However, discharge is only achieved through R2. The ratio of the two resistors sets the frequency of oscillation and this should not be confused with the more usual arrangement of the astable multivibrator in discrete component form described earlier. The capacitor charges and discharges between two voltage levels, i.e. $1/3\ V_{cc}$ and $2/3 V_{cc}$. Consequently if a ground reference is required this will need to be clamped and is effectively done by the internal bistable circuit. Frequency is totally independent of supply voltage variation. The component values can be found by using the chart in Fig. 66d. Notice that the graph reference lines represent the combined values of the two resistors, i.e. $R_1 + 2R_2$; therefore their individual values must be interpolated on this basis. This chart can be used for the following applications as well, even when R_2 is a 'short-circuit' (zero ohms).

Fig. 66. 555 astable multivibrator

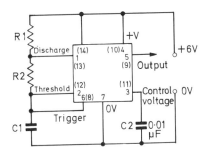

Fig. 66a. The single timer (555) in an astable multivibrator arrangement. Pins are shown for the dual version 556; those in parentheses for the second timer

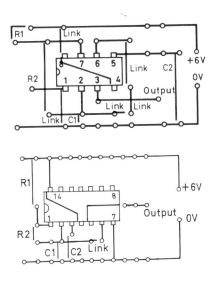

Fig. 66b. Suggested layout and connections of the 555 8-pin DIL package

Fig. 66c. Layout and connections of half of the 556 14-pin DIL package

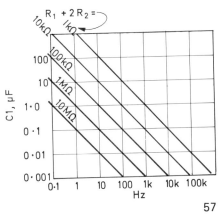

Fig. 66d. Chart showing the relationship between the timing component values C1, R1, R2 and frequency

57

Missing Pulse Detector

Where a steady train of pulses is provided by a pulse generator or astable multivibrator for timing calibration or sequential switching,

Fig. 67. Missing pulse detector

Fig. 67a. The circuit of a 'missing pulse' detector

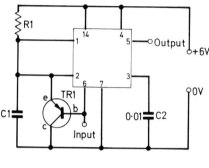

Fig. 67b. Connection diagram of the missing pulse detector circuit using half of a 556

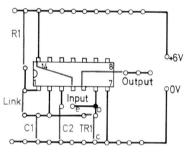

the absence of one pulse, due possibly to a fault condition or interference from an obstruction in a moving conveyor system, can be detected by the 555 timer. The circuit arrangement for this application is shown in Fig. 67.

Pulse Width Modulator

By using the earlier circuit of the free-running multivibrator as a 'clock pulse' generator, the other half of the 556 can be wired as a modulator for use in radio control circuits. The second timer circuit is triggered from the first timer and the threshold voltage is modulated by the signal applied to the control voltage terminal.

The complete layout of the free-running multivibrator and the modulator using one 556 i.c. is shown in Fig. 68.

Fig. 68. Pulse width modulator

Fig. 68a. Circuit of the pulse width modulator with input and output waveforms

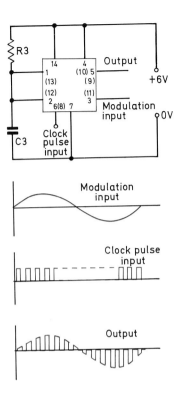

Fig. 68b. Connections of a dual timer 556 for the complete pulse width modulator, including the clock pulse generator

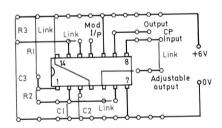

Pulse Position Modulator

By connecting a triangular waveform modulation input signal to pin 3 of the free-running multivibrator, the pulse position is changed with variation of this signal. This is caused by a variation of threshold voltage so altering the timing.

This circuit involves only a simple modification of the layout of the free-running multivibrator.

The dual timer layout (556) is easily wired up to produce a train of pulses triggered by a monostable or 'one-shot' multivibrator. By using one of the timers for this and the other wired as an oscillator, a useful controlled tone burst circuit is available and can be used again for radio control applications or tone-burst recording on tape. It is, in fact, a modified form of pulse width modulator. The circuit and layout is shown in Fig. 69.

Fig. 69. Pulse position modulator

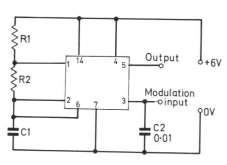

Fig. 69a. Modified astable multivibrator to accept a modulation input

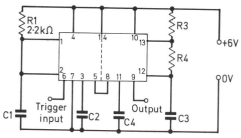

Fig. 69b. Dual timer 556 connections to form a complete 'pulse position' modulator

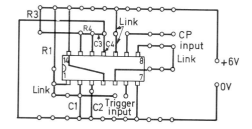

Fig. 69c. Connections of a dual timer 556 for a pulse position modulator

Clock Pulse Generator for TTL Logic I.C.s

A clock pulse generator is often required for use with TTL logic integrated circuits. This is easily done if the input and output of a 555 timer are coupled to TTL gate circuits. Fig. 70 shows how this can be done with two 2-input gates in a 7400 type i.c. so giving what is called a 'correct interface'. As a complete module, this circuit is triggered by the logic information at the inputs of the first gate. This is a NAND

Fig. 70. The timer i.c. and TTL

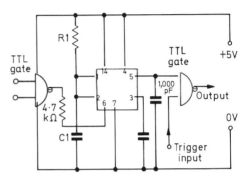

The timer i.c. can be linked to TTL logic i.c.s by using two TTL gates as shown here. Pin connections are for half the dual version 556

gate, so a '1' pulse applied to both inputs will provide the 555 with a '0' at pin 6. The output from pin 5 sets the output gate ready to operate in conjunction with an external applied pulse. The 555 8-pin i.c. layout can be used with the 7400 gates.

In using the 555 or 556 timer i.c., a wiring board or printed circuit is easily rationalised; again one common printed circuit pattern can be made up for several applications.

Fig 70a.
Pin connections for the 555 single timer in 8-pin dual in line package

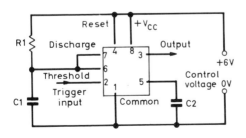

5 D.C. power supplies

All too frequently, d.c. power supply circuits tend to be relegated in the constructor's mind to a position of little significant importance, but modern circuits, particularly those using integrated circuits, can give rise to problems unless the constructor is prepared to spend some thought on the power source.

Traditionally with semiconductor circuits, a dry battery is a convenient and simple supply source. For experimental work this is fine until it runs down and exhibits annoying effects and unreliable circuit operation. The problem with dry batteries is that ageing manifests itself in a change of internal electrical conditions that can have an effect on the performance of external circuitry.

Fig. 71. Battery power supplies

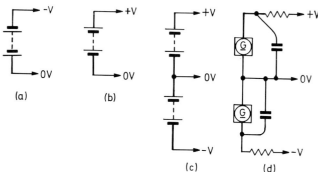

Circuit representation of battery power supplies. (a) Negative line (pnp); (b) Position line (npn); (c) Positive and negative lines; (d) Theoretical equivalent of (c) with the internal battery resistance shown. The capacitance storage effect is also shown

The battery has internal resistance which increases with age and use. Fig. 71 illustrates the electrical circuit of a battery with polarity notation commonly found in electronic circuits. The most often used polarity is with the negative terminal given zero voltage or 'earth' reference but there are cases when both positive and negative voltages are required with respect to a zero reference. The basic theoretical equivalent is shown on the right and included in each battery is a resistive factor dependent on the electrolyte and state of charge stored. The capacitor symbol indicates the storage effect which, in conjunction with the load, acts as a timing circuit, just like a CR network. The rate of discharge depends on the load resistance.

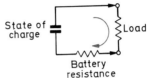

Fig. 72. Battery and load

Theoretical circuit of a battery connected to a load, in terms of capacitance and resistance

Rearranging this circuit as in Fig. 72 we can see that the internal resistance of the battery must be taken into account when assessing the useful state of charge at a particular moment. If the battery has deteriorated its resistance will have increased; the load resistance applied to the terminals will have an effect on the voltage available to it, the more so when battery resistance is high through being run down.

Consequently a constant voltage can never be made available at all times from the battery and this will affect the efficient operation of semiconductor (or valve) circuits being so powered. A typical example of this effect is the severe distortion of battery operated radio receivers when the battery is run down.

To obtain a constant voltage supply it is best to use a power supply unit that can be operated from the a.c. mains. This would seem to be easy enough because the mains voltage remains sufficiently constant (or nearly so) as to have little or no detrimental effect on circuitry being driven. Remember that the most important feature of the power supply, apart from being capable of providing the power required, is that the voltage at the terminals should remain constant over the full range of loading conditions that it is designed to handle.

Transformation

The first step is to convert the mains voltage to a voltage nearer to that required, allowing a little extra in hand for a small degree of loading.

Fig. 73 shows that this is done by using a voltage transformer designed to convert a.c. mains (240 V r.m.s. nominal in the U.K.; 220 V r.m.s. in Europe, 110 V r.m.s. in the USA) to a low r.m.s. voltage. As an

Fig. 73. A.C. power supplies

Fig. 73a. Half-wave rectifier supply

Fig. 73b. Relationship between a.c. voltage levels

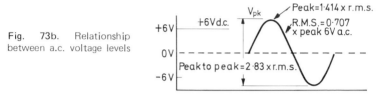

example, let us assume that 6 V d.c. will be required ultimately; the recommended a.c. voltage from the transformer can be 6 V r.m.s. Fig. 73b shows a graphical representation of these in relation to each other. In actual fact the r.m.s. voltage is an approximation only of the near equivalent d.c. voltage from a practical point of view, having similar electrical effects on a resistive load at that voltage. For capacitive or inductive loads the effects are noticeably different.

Rectification and Smoothing

To convert the a.c. voltage to d.c. a rectifier must be used; this can be simply a single diode, a double diode or a bridge rectifier. Each is represented in Figs. 73 and 74 by the appropriate symbol together with the output waveform.

What has been achieved here is to make the available current flow in the same direction through the load and prevent it flowing in the opposite direction. However, the voltage still varies so we must smooth it out to a uniform level. By connecting a high value capacitor across the output from the rectifier the resulting voltage across the load will improve a little as shown in Fig. 75.

In Fig. 75a, the voltage waveform is still rather 'lumpy', but is better in (b) and (c). Capacitors, like batteries, have an internal resistance which influences a.c. or fluctuating d.c. voltage, therefore to reduce this effect to as low a level as possible the capacitance value

Fig. 74. Full-wave rectified power supplies

Fig. 74a. Using two diodes;
d.c. voltage obtainable $\cong$
0.9 r.m.s. voltage

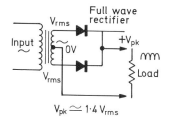

$V_{pk} \cong 1\cdot 4\, V_{rms}$

Fig. 74b. Using four diodes
or a bridge rectifier module

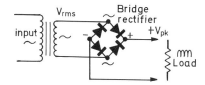

Fig. 75. Smoothing

Adding a reservoir capacitor to assist the positive going voltage in keeping high for each cycle of a.c. input

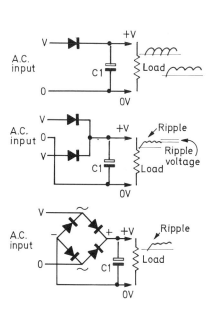

must be as high as is practicable. It will be charged by the applied voltage and discharged via the load and power source. To minimise the effect of the power source on loading the capacitor, the combined resistance of the transformer winding and rectifier must be kept as low as possible. The bridge rectifier arrangement is the best to choose from this point of view because it helps to lower the effective overall resistance.

65

In spite of all these precautions the output voltage is liable to fluctuate. It is generally referred to as ripple, which should be reduced to an absolute minimum by means of further filtering, using a CR or LC network as shown in Fig. 76. The second capacitor also acts as a secondary reservoir or store and further reduces the source resistance presented to the load. The mean output voltage is approximately 1.4 times the r.m.s. voltage, measured across the first capacitor.

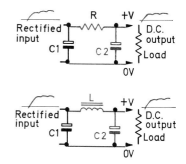

Fig. 76. Extra smoothing

Extra smoothing and reduction of the ripple voltage by adding an RC or LC filter. C2 helps to lower the source impedance as well as shunting the ripple frequency

Fig. 77. Decoupling

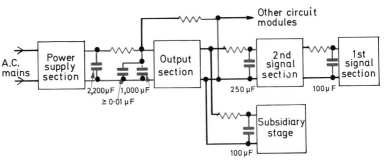

An example of how decoupling is applied to power supply lines, in order to prevent voltage deviation at one stage, due to loading or transient interference from having an adverse effect on another stage carrying small signals. The capacitors 'decouple' such fluctuations by presenting a low resistance to a.c. The resistor values will be based on the current and voltage requirements of each section

This is the basic principle behind the simple a.c. mains power supply. It is the minimum that should be used, although it does not take any account of variations in mains supply voltage, nor will it be immune to the effects of variable loading conditions beyond a few milliamperes. Some circuitry requires stringent mean voltage regulation to avoid the possibility of unwanted modulation of the d.c. supply by interfering signals. Further decoupling networks (Fig. 77) are frequently used in multi-stage equipment to minimise this latter effect, for example, in consumer electronics.

Of course, in designing power supply requirements allowance must be made for inevitable voltage drop across each of the decoupling resistors. If one stage requires relatively high power, it can cause the supply voltage to change under heavy loading conditions and consequently upset the voltage to the next stage unless the decoupling networks are used.

A suggested layout for this basic power supply is shown in Fig. 78. The components can be mounted on the transformer or be made part of the complete equipment assembly board. The layout shown allows for any of the three types of circuit shown to be assembled; further decoupling will often be accommodated on the equipment assembly

Fig. 78. Simple power supply

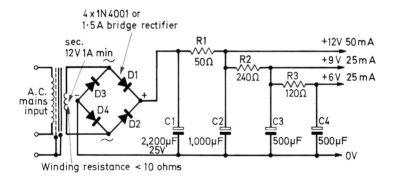

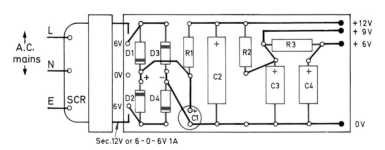

Circuit diagram and suggested layout of a simple power supply. Black line connections can be made as tinned copper wire or converted to a printed circuit board layout. All resistors should be 1W types mounted clear of the board

but can be incorporated on this board if preferred. An important consideration in determining the layout design will be the physical size of the components used, especially capacitors. The example given is for a 6 to 12 V d.c. supply.

It is always recommended to use as large a transformer as is practicable, with an in-built screen between windings, and connected to 'earth', to minimise the effects of interwinding capacitance and mains borne interference. The accuracy of the voltages at the terminals will depend on the consistency of load currents and the secondary winding resistance of the transformer. Minimum winding resistance is desirable but the lower this is, the larger will be the physical size of the transformer. The current rating will also be higher.

Whilst giving adequate performance at relatively low current levels, up to those given, there are likely to be shortcomings when applied to projects using logic i.c.s or op-amps. The following pages show the development from this basic design to more sophisticated circuits that are regulated against supply or load fluctuations. Added refinements are explained as we come to them. By using a bridge rectifier the output is 'floating', therefore either positive or negative can be made 'earth'.

Voltage Regulation

The simplest form of voltage regulation for supplies that are likely to be subject to small variation, consists of a zener diode. These useful devices will hold the line voltage down to a steady value. They are not intended to, and often not capable of, effecting a large voltage drop. The zener is connected across the supply lines as shown in Fig. 79a for a 6 V supply. Fig. 79b shows another circuit which includes provision for a 5 V output to supply DTL and TTL logic i.c.s. The low value capacitance C3 across C2 compensates for the inability of the large capacitor to suppress fast voltage spikes on the supply lines, which can otherwise cause spurious triggering of logic switching circuits. A ripple voltage better than 50 millivolts (0.05 V) can be expected with these two circuits. A suggested layout is given in Fig. 79c. If the 5 V output is not required, the appropriate components at the right-hand end can be omitted.

If the ripple voltage is an important consideration with constant voltage under heavier loading conditions, then it is well worth considering the addition of a stabilising transistor. The extra cost will be justified in the long run. A 6 V basic stabilised supply circuit is shown in Fig. 80. For a 9 V output at 250 mA, change the transformer to

Fig. 79. Regulation of power supplies

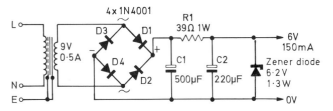

Fig. 79a. Using a zener diode to regulate the output voltage

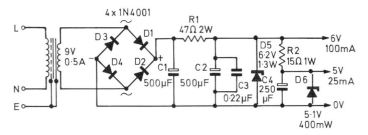

Fig. 79b. Two output voltages with zener regulation

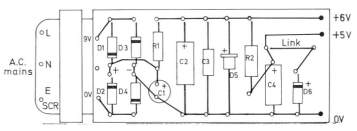

Fig. 79c. Suggested layout of the circuit shown in Fig. 79b. The layout is similar to that for the 12 V supply but with two more component positions

12 V secondary 0.5 A minimum and the zener diode to a 9.1 V type. Output ripple voltage for this kind of stabilised circuit should be better than 25 mV at 250 mA. The stabilising transistor is an AD161 rated at 4 W and should be mounted on a metal plate heat sink.

Overload Protection

If there is a risk of short-circuit occuring across the supply lines, either through accident or breakdown there is a distinct risk of damage to components within the power supply unless some kind of protection circuit is included. The circuit in Fig. 81 shows how this can be done for a 6 V supply. A simple comparator circuit detects a threshold voltage which is compared with that set by the potentiometer VR1.

Fig. 80. Stabilisation of power supplies

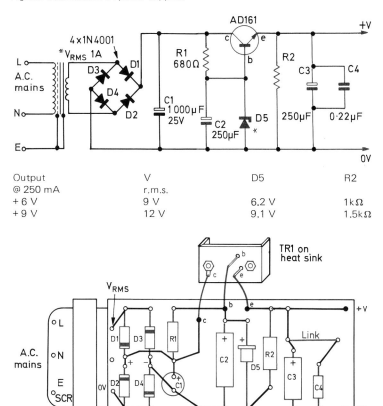

Output @ 250 mA	V r.m.s.	D5	R2
+6 V	9 V	6.2 V	1kΩ
+9 V	12 V	9.1 V	1.5kΩ

Using a power transistor to help stabilise the output

As the load current rises above the designed maximum, the stabilising transistor is used as the threshold compensator acting as a negative feedback path to the supply. The differential pair of transistors should be perfectly matched, or may be substituted by an operational amplifier i.c. such as the 741. This particular circuit can continue to operate under short-circuit conditions without damage.

Fig. 81. Short-circuit protection

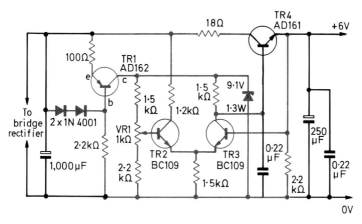

Stabiliser circuit with some measure of short circuit protection provided by the differential amplifier. The mains transformer should supply 12 V r.m.s. to a bridge rectifier made up with four 1N4001 diodes

Another form of short-circuit protection is shown in Fig. 82. Here the threshold detector applies direct negative feedback bias to the stabilising transistor, acting as a base current regulator. In both of these

Fig. 82. Overload protection

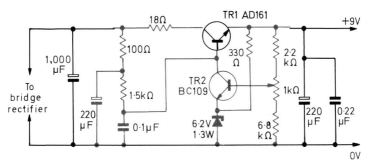

Another stabiliser circuit using a simple transistor current amplifier in negative feedback mode to protect against overload

circuits, the protection arrangement also provides self-compensation for small load current variations, so giving a much improved ripple voltage better than 10 mV. The load current rating is 250 mA.

Where circuits require both positive and negative supplies, two stabiliser circuits can be used but it is more sensible and economic to arrange both from a common transformer. Fig. 83 shows some of The ways that this can be achieved. Fig. 83a shows how a common

Fig. 83. Dual power supplies

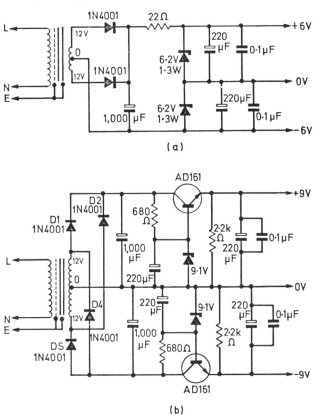

Dual power supplies from one transformer and full-wave rectifier diodes

positive supply can be converted for both positive and negative. This uses two zener diodes that are connected in series, the centre-tap being the zero voltage reference or earth. A centre-tapped transformer with full-wave rectification is used but a bridge rectifier is equally suitable. It is important not to connect the transformer secondary winding to earth. Voltage control is held down by the 6.2 V zener diodes. Where a stabilised output is required the circuit in Fig. 83b can be used. Heat sinks should be provided for the transistors. The output ripple voltage can thus be reduced from about 300 mV to about 20 mV. The secondary winding of the transformer should have a d.c. resistance of less than 10 ohms. Short circuit protection can be provided by using a sensing and control transistor to regulate the base current of the stabilising transistor in each section. The overload current is detected by a fall in the voltage across the potential divider network; this operates the complementary pair of control transistors. The voltage at each output is set by the preset potentiometers VR1 and VR2.

6 A.C. amplifiers

We have seen from the preceding chapters how d.c. voltages and current can be processed in electronic circuits. These are particularly applicable where switching, measurement and control are required. The other kind of electronic circuits that are probably most commonly used are those that process a.c. signals. These include all forms of entertainment electronics: radio, television, recording, electronic music, and so on.

To process an a.c. signal, a transistor must be made to operate so that the input signal will be faithfully reproduced at the output, but perhaps at a higher voltage or current. At the same time the transistor must not be allowed to run at such high levels as to cause overheating, and hence thermal runaway damage, to the crystalline structure that makes up the operative part. Manufacturers' data indicate safe operating current and voltage levels, although precise characteristics will vary from one specimen to another of a particular type. In particular, current gains quoted will often fall within a wide tolerance range and will vary according to the collector current and applied voltages.

To keep the circuit conditions within the limits specified it might seem an easy matter just to apply specific d.c. voltages and the constructor would be excused for assuming that everything would be theoretically satisfactory. Thermionic valves are extremely tolerant of overload conditions and are unlikely to exhibit instant breakdown symptoms unless physical distortion of the electrodes is caused by overheating. Since the heater element would break down first, this situation is not likely to arise.

Transistors and, indeed, any other semiconductor devices, depend for their thermal stability on external components to limit their current flow. If there is a risk of violent or high power fluctuations in these conditions, then simple added protection can be provided. It is a little like keeping an engine cool by applying a water jacket or blowing air onto it. The semiconductor device sometimes needs that added safeguard to allow excessive heat to be taken away from it when the going gets temporarily difficult to handle.

The need for additional protection from overheating is usually only necessary in situations where high power or current is handled; if the device cannot cope easily at lower power, then obviously a higher rating device must be substituted. In all cases the applied voltage and currents must be reasonably controlled.

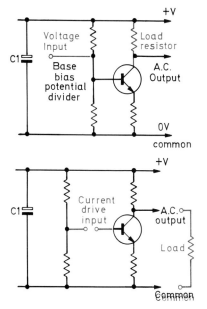

Fig. 84. Base bias

A single transistor amplifier stage with base bias potential divider

Fig. 84a. Voltage drive input and collector load resistor

Fig. 84b. Current drive input and external loading. Capacitor C1 effectively makes the positive supply line equal to the common line for a.c. signals so that the load may be connected to either line

There are several textbooks that explain semiconductor a.c. theory in a working circuit. It will be found that, above all, collector current and base voltage must be kept within reasonable limits for the device in hand. The most important factors, then, are the base bias setting and the amount of current, determined to some extent by the series emitter resistor in a.c. applications (Fig. 84).

Base Bias Stabilisation

A.C. signals are applied to the base (in common-emitter mode) as a fluctuating current. The collector current will be, under correct operating conditions, roughly proportional to it. Because a.c. is applied, the transistor recognises the signal as a fluctuating d.c. but interprets this at the output as a.c. This is where voltage reference points become important because a.c. consists of negative-going as well as positive-going voltage and current. The zero reference for d.c. conditions is assumed to be the emitter. The base is, therefore, made to swing in negative and positive magnitude with reference to the emitter. The emitter resistor must be capable of providing sufficient d.c. voltage drop across it so that the negative-going swing at the collector reaches its full potential.

Fig. 85. A.C. signals applied to a transistor amplifier

Fig. 85a. Grounded emitter reference, no emitter resistor

Base and collector positive going voltage

Fig. 85b. Emitter resistor in circuit

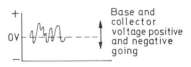

Base and collector voltage positive and negative going

Fig. 85c. High amplitude peak, negative going voltage distorted

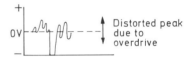

Distorted peak due to overdrive

Fig. 85 shows the various waveforms that are represented by this reference. The a.c. input signal at the base will vary in a positive and negative going manner with reference to the emitter.

It is worth making up a simple circuit that will show these points and if you are fortunate to have access to an a.c. signal generator and an oscilloscope, you can view the effects by looking at the output waveform, first with no emitter resistance, then with a resistance, and by increasing the voltage of the a.c. signal from the generator, observe at what point distortion begins to occur due to the inadequacy of the bias resistance. Remember that the higher value of emitter resistance might permit a greater voltage swing, but it also limits the current through the collector and load resistance thereby limiting the output. Less current will also flow though the transistor so keeping overheating problems down, but at the expense of lower gain.

Most a.c. amplifiers are based on this principle and further extended to other a.c. circuits such as oscillators. The emitter resistance will not always be used because there are other a.c. waveforms (square, sawtooth, etc.) that are recognised by the transistor as modified d.c. levels that can be operated with a different zero reference level. Where sine waveforms are processed, such as in audio and radio frequency application, the emitter reference must be maintained to avoid squaring distortion.

D. C. Feedback Stabilisation

Bias stabilisation as shown in Fig. 84 is satisfactory for an individual stage; in multi-stage circuits this approach can be extravagent on components required and even cause slight deterioration of a.c. performance. Fig. 86a shows an alternative circuit that uses d.c. feedback stabilisation so that direct coupling of the transistors can be achieved. Notice that frequency sensitive components are thus minimised. The

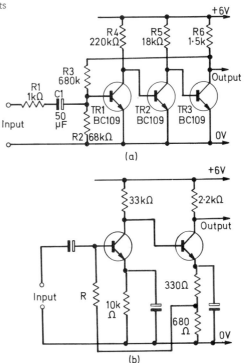

Fig. 86. D.C. feedback circuits

Two methods of providing d.c. feedback stabilisation

voltage gain of this circuit is about 500 for a uniform frequency response from 10 Hz to 20 000 Hz. Notice, however, that there are no emitter resistors so outside this frequency range a square wave output waveform would become distorted. D.C. bias stabilisation is provided over three stages from the output to the input. This is set by a 680 kΩ resistor and if drastically altered will lead to a risk of instability and possibly transistor thermal runaway. The output impedance is about 1kΩ. The input impedance, determined by R1, is also 1kΩ.

Because of the use of direct coupling, this circuit is extremely simple and can easily be made up on a strip-board, Veroboard, or perforated board with copper wire. Fig. 87a shows a suggested layout and this can be interpreted into a p.c.b. form if required. A position is also provided on this diagram for input and output coupling capacitors that would be necessary.

Fig. 87. D.C. feedback layouts

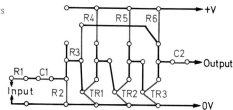

Fig. 87a. Suggested wiring layout of the circuit in Fig. 86a

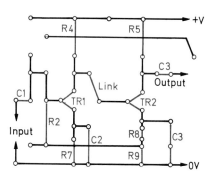

Fig. 87b. Suggested wiring layout of the circuit in Fig. 86b

Variations in emitter circuitry for an amplifier stage can take the form shown in Fig. 86b. The emitter resistor is shown and provides what is called a.c. feedback, that is, it creates a current carrying path for a.c. signals which will reduce the amplifier gain significantly. To offset this loss, a large value capacitor is connected across it so creating a short circuit path from emitter to ground for a.c. signals, whilst the d.c. conditions remain unaffected. Sometimes it is useful to keep just a small amount of this kind of feedback in circuit to help stabilise the a.c. operating conditions (Fig. 88).

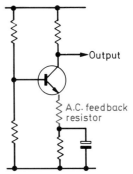

Fig. 88. A.C. feedback

Common emitter amplifier with a.c. feedback resistor in the emitter circuit

Two stage d.c. feedback bias can also be taken from a potential divider in the emitter of the second transistor back to the base of the first (Fig. 86b). Modifications to the wiring layout can easily accommodate these various alternatives as suggested in Fig. 87b.

Layout

These are only a few possible amplifier stage arrangements. In building high quality amplifiers for sound reproduction, radio or television applications, it is likely that the constructor will have to cope with special layout requirements to avoid problems of instability, stray capacitance effects, cross-modulation distortion and even radiated interference. Each design has its own special characteristics; it does not necessarily follow that these problems are inevitable. In designing a circuit layout the design applications engineer will have to anticipate these problems, carry out a basic first layout, then modify it as necessary.

When working on printed circuits, conductor re-routing, component position change and maybe electrostatic screening will be tried to overcome symptoms of undesirable spontaneous circuit behaviour. The same applies to wiring of components on a plain perforated board. In fact, it is often a better plan to build up a first prototype layout using this method. It is quicker and much easier to make alterations. Thin tinned copper wire is suitable and the components can be mounted by passing their wires through the holes of the board and wiring underneath.

It is not possible to generalise on procedures for these applications; it will largely be as a result of experience that the constructor will recognise the problems and how to avoid them. There are, however, a few guidelines that all beginners should learn as being basic to practically all circuits. These are summarised as follows:

1. Keep the first stage (or first transistor) at one end of the layout and the last stage (output) at the opposite end. The intermediate stages will fall naturally into sequence between these.
2. Where two or more identical channels are to be constructed, keep them side by side, adopting identical layouts and using identical components for each.
3. Try to arrange for external connections to be made at or near the edge of the board nearest to the point of application. Inputs and outputs should be kept separate.
4. Stages that process very low voltage signals are most vulnerable to external electromagnetic and electrostatic influences; keep them away from power amplifier stages, power supplies and a.c. carrying wiring. If necessary use screened cable that has a very low capacitance characteristic (centre-core-to-screen) particularly for high impedance circuits.
5. Keep all wires as short as possible, particularly in radio and television signal stages and other circuits operating at very high frequencies.
6. Observe the guidelines given for printed circuit design (see the Constructor's Guide *Printed Circuit Assembly*) regarding conductor size and spacing, and the use of a large area 'ground plane' around the perimeter of the board.
7. Allow for the use of heat sinks for power amplifier and power stabiliser stages and allow plenty of space for air circulation. Keep these away from other semiconductors that are vulnerable to high ambient temperature problems.
8. Avoid the use of excessive crossover connections and link wires. Much of the trouble that can arise in layout design can be averted by using integrated circuits.

The integration of component elements within the i.c. package is such that long connections are avoided, thermal stability is improved and radiation effects are virtually non-existent. This would seemingly imply that integrated circuits are the best method to use. In many cases this is so, but experienced designers have found that there are performance limitations in integrated circuit amplifiers that restrict their usage to non-critical applications. Integrated circuits are also limited in their power handling capability compared with some discrete transistors.

Choice of I.C.

In choosing a suitable i.c. it is most likely that a few additional components will be needed to adjust its performance according to requirements, within its capability. Some examples of i.c. amplifer circuits are shown in the following pages, the i.c. being represented by the customary triangular symbol. The numbers are the pin connections for inputs, outputs, power supplies and filter networks. Volume and tone controls, tuned circuits and decoupling capacitors will probably be needed as suggested.

The main advantage of using integrated circuits here is that a very much simpler printed circuit board or wiring layout can be adopted. To enable alterations to be made or to facilitate easy removal of the i.c., it is recommended that i.c. holders are used where possible; permanent soldering of the i.c. itself can be carried out if the constructor is satisfied that the complete circuit is working satisfactorily.

CA3052

One example of the use of an RCA integrated circuit in an amplifier is shown in Fig. 89. The CA3052 contains four individual matched small signal amplifiers which are shown connected as a stereo preamplifier unit. The equalisation circuits will be designed to individual requirements and the components used are mounted on the same board as the i.c. Each of the amplifiers is an op-amp with characteristics suited to audio applications.

There are several integrated circuit amplifiers available apart from the many operational amplifiers described earlier. Of particular interest to many constructors will be those that provide a complete packaged

Fig. 89. CA3052 i.c.

The CA3052 used as a stereo preamplifier/control unit. Second channel pin connections are shown in parentheses

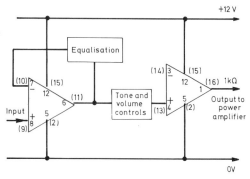

81

audio amplifier with output stages included, since these are very useful for quick assembly, experimental work and providing inexpensive audio amplifiers for record players, radio and television receivers. They have facilities for providing your own tone control networks and feedback or a.g.c.

Basically all of these i.c.s use the differential amplifier, as do op-amps, because of the superior stability under fluctuating environmental conditions. However, they are still vulnerable to damage from overheating or excessive load currents, unless a built-in protection circuit is included. If additional power output is required they can often be connected to high power transistor output stages, but allowance must be made for the extra current required from the power supply.

For general purpose applications, there are several i.c.s, some of which are shown here. Power output is usually up to about 20 W although most are less than 10 W (r.m.s rating). Special types rated at higher power output usually consist of either a packaged module with discrete components incorporated or are developed as 'thick-film' circuits.

TBA651

The TBA651 (Fig. 90) is a popular i.c. that is used in a.m. radio applications, operating from a supply voltage anywhere between 4.5 and 18 V. It provides an audio signal of up to 0.6 V which can be fed into a suitable power amplifier i.c. It is made in a 16-lead quad-in-line plastic package. The difference between this and the dual in-line package is that alternate pins of the quad-in-line package are bent into staggered array so making printed circuit board design easier to accomodate the close connections.

At first, the constructor might be bewilderd by the numerous connection descriptions, but as in any application of an integrated circuit it is most important to consider it as a 'black box'. Only the basic active circuitry incorporating transistor amplifiers is included within the package; it is left to the constructor to add the external components to complete the circuit.

Although seemingly contradicting the philosophy of integrated circuits, this method does leave some flexibility of application because the external component values can be chosen to suit. Unfortunately there are not sufficient data provided in most cases to assist the constructor in his choice of component values so he is well advised to study circuits published by the manufacturers or elsewhere in magazines.

Fig. 90. TBA651 i.c.

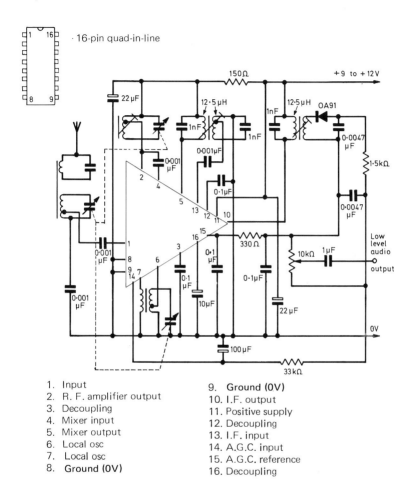

16-pin quad-in-line

1. Input
2. R. F. amplifier output
3. Decoupling
4. Mixer input
5. Mixer output
6. Local osc
7. Local osc
8. Ground (0V)
9. Ground (0V)
10. I.F. output
11. Positive supply
12. Decoupling
13. I.F. input
14. A.G.C. input
15. A.G.C. reference
16. Decoupling

The TBA651 in a superhet a.m. receiver circuit. Tuned circuit components will depend on the frequency of operation

The data for the TBA series is taken from the SGS-Ates integrated circuit data handbook and for the CA series from the RCA handbook. These and others from various manufacturers have been developed subjectively with specific applications in mind, so it is a sensible plan to use the information provided in conjunction with electrical characteristics and suggested constructional notes.

Practical winding data for the tuned circuits is not supplied but the experienced constructor will be able to make his own according to the receiver frequency range. Less experienced constructors are advised to use standard ready made tuned coils. In this case they would be the r.f. and i.f. standard coils for medium wave and long wave reception. The parallel capacitors should be selected accordingly.

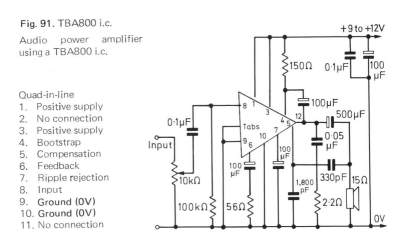

Fig. 91. TBA800 i.c.
Audio power amplifier using a TBA800 i.c.

Quad-in-line
1. Positive supply
2. No connection
3. Positive supply
4. Bootstrap
5. Compensation
6. Feedback
7. Ripple rejection
8. Input
9. Ground (0V)
10. Ground (0V)
11. No connection

TBA800 and TBA810

Two popular amplifiers are the TBA800 and TBA810, both provide a medium power output when used with a suitable heat sink. Without the heat sink, power is limited to less than one watt under reasonably safe heat conditions. However, audio amplifiers are often called upon to handle very large instantaneous peaks of high power for very short periods. The i.c. must be capable of handling these peaks without risks so heat sinks are always recommended.

Supply voltage for power amplifiers is often higher than in small signal amplifiers in order to help provide the audio power output. A stabilised power supply is recommended for stable operation, but not essential. Class B arrangement is usual for the output stages and in most cases a series output capacitor will be needed to isolate the loudspeaker from the d.c. supply.

In making a heat sink for both the TBA800 and TBA810 the wide tabs on the package can be carefully soldered to a sheet of copper bent to the shape shown in Fig. 92, or by soldering to a large area of printed circuit board. Make sure that the heat sink is insulated from the pins.

Fig. 92. Heat sink

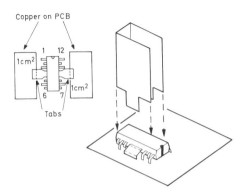

Soldering should be completed within 10 seconds to avoid overheating and damage to the circuit. Recommended heat sink area for these is based on a temperature coefficient of 30°C temperature change per watt and should be at least 2.5 mm^2 for a thickness of about 0.5 mm. It should be painted matt black after assembly for maximum heat dissipation.

Fig. 93. TBA810

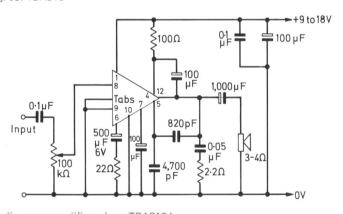

Audio power amplifier using a TBA810 i.c.

The published data on these two devices is as follows:

	TBA 800	TBA 810
Power supply recommended	24 V	16 V
Quiescent current (pin 12)	9 to 20 mA	12 to 20 mA
Bias current (pin 8)	1 to 5 A	0.4 A
Output power	5 W	7 W
Input voltage	220 mV	220 mV
Input sensitivity	80 mV	35 mV
Frequency response (−3dB points)	40-20 kHz	40-20 kHz
Distortion (up to 2.5 W)	0.5%	0.3%
Efficiency	70%	70%
Load impedance	16	4
For low supply voltage connect resistor between pins 1 and 4	150Ω	100Ω

Characteristics in full and design graphs are shown on the manufacturer's data sheets.

TDA2020

Of more recent development is the TDA2020, a nominal 15 W integrated circuit amplifier that is capable of giving up to 20 W peak. The performance of this i.c. is made possible by applying a positive and negative d.c. supply of up to 18 V; total harmonic distortion at 15 W is 1%. The load is direct coupled, no capacitor is needed. The secret behind this achievement is the fact that the chip is soldered directly to a copper heat sink within the 14-pin DIL package. This amplifier can also be used as a power op-amp delivering 2 A output. The specification is as follows:

Nominal supply voltage	plus and minus 15 V
Quiescent current for supply voltage of 15 V	50mA
Input current	1 μA
Open loop gain (no feedback)	90 dB
Input impedance	1 MΩ
Load impedance	4 Ω

A heat sink in contact with the copper surface is recommended. The device incorporates 'power limiting' and thermal protection. A suggested circuit with external components is shown in Fig. 94.

Fig. 94. TDA2020 i.c.

Audio power amplifier using a TDA2020 i.c.

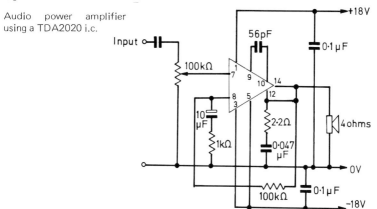

Fig. 95. CA3090

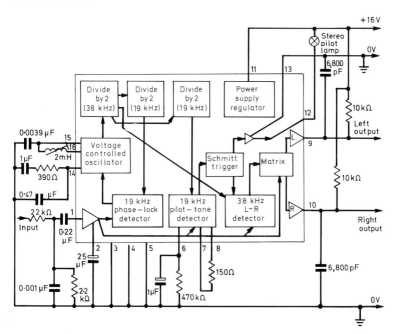

Block diagram of the RCA stereo multiplex decoder i.c. CA3090 with associated external components

87

CA3089 and CA3090

Two particularly popular integrated circuits are the CA3089 and CA3090 which are used in high quality f.m. radio tuners. The CA3089 is the i.f. amplifier section and the CA3090 the stereo decoder. Fig. 95 shows a block diagram of the internal details of the stereo decoder i.c. with suggested external components added. It provides automatic stereo switching, operating from a power supply of 16 V. An internal voltage regulator permits satisfactory decoding over a wide range of supply voltage, only one coil being required. Printed circuit layout is critical, so it is recommended that the constructor should follow the design published in the manufacturer's data sheet or that given in the specific project details published or provided for a complete stereo tuner.

Similarly with the CA3089E f.m. i.f. system, which is shown in block diagram detail in Fig. 96. A three-stage i.f. amplifier is included with a level detector for each; a balanced quadrature f.m. detector and audio amplifier are also included. The external components shown include a tuning meter and connection can be made to a.g.c. and a.f.c. circuits. Ceramic i.f. filters can be used in place of the 10.7 MHz tuned circuit. The audio output pin 6 would be linked to the input of the multiplex decoder CA 3090.

Fig. 96. CA3089E

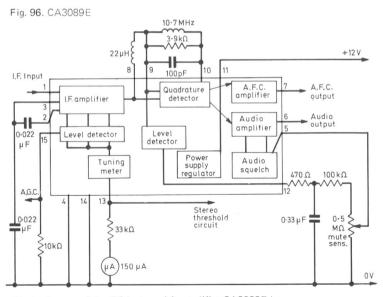

Block diagram of the RCA stereo i.f. amplifier CA3089E i.c.

Appendix

Component Selection

The circuits shown in this book use common readily available components. Constructors who would like to try making up any of these may find the following components lists useful. The style of presentation is similar to that used by the popular constructional magazines published in the U.K. Constructors in other countries should verify correct translations to avoid wiring the wrong components. For further explanations on the identity, ratings and usage of components readers are recommended to refer to the Constructors Guides on *Electronic Diagrams* and *Electronic Components*.

The components shown in the circuit diagrams in this book have reference code numbers for identity only in these applications, for example, resistors R1, R2, R3, etc. These have no bearing on the function of the components. The following lists quote the component ratings which must be quoted when ordering.

Whilst every effort is made to avoid errors, neither the author nor the publishers of this book can be held responsible for any malfunction of circuits made up by the reader. Constructors are urged to take due care and heed the advice given in the books in this series, especially with regard to soldering.

In order to simplify and clarify the components lists, abbreviations have been used in line with common practice in the U.K. These are explained elsewhere in the series and include the following:

Resistors (R)

All 0.5 watt carbon film except where stated.

Ω ohms   kΩ kilohms,   MΩ megohms

Capacitors (C)

All rated at battery supply voltage given or higher, except where stated.

    pF picofarads,   nF nanofarads,   µF microfarads

Others

    V volts,   mV millivolts,   W watts

    A amperes   mA milliamps   µA microamps

In many cases additional parts may be required, such as tag strips, copper clad strip-board or printed circuit board for mounting the components. The constructor should have an adequate stock of nuts and bolts, solder and other hardware. A number of short connecting leads will be useful with crocodile clips on the ends for easy connection of batteries, meters and loudspeakers, etc.

Constructors can obtain these components from retail outlets or by writing to Messrs. P.C. Services, 282 Hatfield Road, St. Albans, Herts. A stamped addressed envelope should be enclosed if a reply is required. Printed circuit boards and Veroboard are also available; the Figure number of the circuit must always be quoted.

Components Lists for circuits shown in this book.

Fig. 8, page 7

Resistors
R1 68kΩ   R2 15kΩ   R3 4.7kΩ   R4 100Ω

Transistors (all npn types)
TR1 BC108 or BC109   TR2 BC108 or BC109

Battery
B1 6 volt type PJ996

Fig. 9, page 9

Transistor
Type BC108 or BC109

Resistors
R1 180kΩ   R2 1.5kΩ

Meter
Multi-range type or separate meters as shown

Battery
6V type PJ996

Fig. 10, page 11

Resistors
R1 180kΩ   R2 1.5kΩ   R3 1.5kΩ   R4 100Ω

Transistors
TR1 and TR2   both BC108 or BC109

Battery
6V type PJ996

Lamp   Switch
LP1 6V 0.06A   Single pole on/off

Fig. 11, page 13

Resistors
R1 100Ω   R2 1.5kΩ   R3 1.5kΩ   R4 100Ω

Transistors
TR1 and TR2   both BC108 or BC109

Lamps
As substitutes for R1 and R4   6V 0.06A

Batteries
6V type PJ996
1½V type SP11

Fig. 12, page 13

Components as in Fig. 11, plus two diodes type 1N4001 or 1N914

Fig. 13, page 14

Components as in Fig. 12 plus:

Resistors   Battery
R5 4.7kΩ   R6 4.7kΩ   B2   3V type 800

Fig. 16a, page 17

Resistors
R1 2.2kΩ   R2 47kΩ   R3 47kΩ   R4 2.2kΩ

Transistors
TR1 and TR2   both BC108 or BC109

Capacitors
C1 and C2   both 0.05μF   50V or more, polyester

Fig. 16b, page 17

Components as in Fig. 11 plus:

Capacitors
C1 and C2   Any values from 10μF to 100μF

Fig. 18, page 21

DTL 2-input NAND gate type MIC949 or made from:

Resistors
R1 2.2kΩ   R2 470Ω   R3 47kΩ   R4 2.7kΩ

Transistors
TR1 and TR2   BC109

Diodes
D1 and D2   Both 1N914 or 1N4001

Battery
4.5V type 1289

Fig. 19, page 22

TTL 2-input gate type 7400

Fig. 22, page 25

Resistor   Transistor
R1 2.2kΩ   Type BC108

Lamp
LP1   6V 0.06A   or
Light emitting diode type TIL209

Fig. 23, page 25

Resistors
R1 2.2kΩ   R2 220Ω

Diode
D1   Light emitting diode type TIL209
or   alternative rated at approx 2V 10mA.

Fig. 24, page 28

Resistors
R1 1kΩ   R2 22kΩ   R3 22kΩ   R4 1kΩ

Transistors
TR1 and TR2   Both BC108 or BC109

Diode   Battery
D1 type 1N914   B1   6V type PJ996

Fig. 29, page 31

Resistors
R1 2.2kΩ   R2 2.2kΩ   R3 15kΩ
R4 15 kΩ   R5 2.2kΩ   R6 2.2kΩ

Transistors
TR1 and TR2   Both BC108 or BC109

Capacitors
C1 and C2   Both 0.01μF, 50V or more, polyester

Batteries
B1 and B2   Both 6V type PJ996

Fig. 30a, page 32

Resistors
R1 1kΩ   R2 1kΩ   R3 3.3kΩ   R4 680Ω

Transistors
TR1 and TR2   Both BC108 or BC109

Capacitors
C1 0.15µF   50V or more polyester

Batteries
B1 and B2   Both 6V type PJ996

Fig. 30b, page 32

Resistors
R1 1.5kΩ   R2 3.3 kΩ
R4 10 kΩ   R5 330Ω   R6 330Ω

Transistors
TR1 and TR2   Both BC108 or BC109

Capacitor
C1 0.022µF

Battery
B1 6V type PJ996

Fig. 31a, page 33

Resistors
R1 6.8kΩ   R2 2.2kΩ   R3 18kΩ
R4 1kΩ   R5 270Ω   R6 2.2kΩ

Capacitors
C1 0.001µF   C2 0.05µF

Transistors
TR1 and TR2   Both BC108 or BC109

Battery
B1 6V type PJ996

Fig. 31b, page 33

Resistors
R1 4.7kΩ   R2 1kΩ   R3 8.2kΩ   R4 18kΩ   R5 1kΩ

Transistors
TR1   BCY71 or BC478
TR2   BC108 or BC109

Capacitors
C1 0.001µF   C2 0.05µF

Battery
B1   6V type PJ996

Fig. 32a, page 33

Resistors
R1 4.7kΩ   R2 1kΩ   R3 10kΩ   R4 10kΩ   R5 1kΩ

Transistors
TR1   BCY71 or BC478   TR2 BC108 or BC109

Capacitor
C1   0.001µF

Battery
6V   type PJ996

Fig. 32b, page 33

Resistors
R1 100Ω   R2 2.2kΩ   R3 2.2kΩ   R4 100Ω

Transistors
TR1 and TR2   Both 2N1671 or 2N2646

Capacitor
C1   0.05µF

Batteries
Two 6V   type PJ 996 in series

Fig. 34c, page 39 (Discrete component version)

Resistors
R1 2.2kΩ   R2 1.8kΩ   R3 4.7kΩ   R4 6.8kΩ

Transistors
TR1 and TR2   Both BC108 or BC109

Diodes
D1 to D5   All 1N914 or 1N4148

Battery
4.5V type 1289

Fig. 34c (Integrated circuit version)

IC1   MIC930 dual 4-input NAND gate (DTL)

Battery
4.5V type 1289

Fig. 37, page 40

IC1 MIC930 (DTL) or 7420N (TTL)

Battery
4.5V type 1289

Fig. 39, page 41

IC1 MIC946 (DTL) or 7400N (TTL)

Battery
4.5V type 1289

Fig. 40, page 41

IC1 MIC935 (DTL) or 7404N (TTL)

Battery
4.5V type 1289

Fig. 45, page 45

Resistors
R1 68kΩ   R2 680Ω   R3 68kΩ   R4 680Ω   R5 1kΩ
R6 1kΩ   R7 47kΩ   R8 47kΩ   R9 33kΩ

Transistors
TR1, TR2, TR3   All BC108 or BC109

Potentiometer
VRI 10kΩ linear skeleton preset type

Batteries
B1 and B2   Both 6V type PJ996

Fig. 46, page 46

Resistors
R1 33kΩ   R2 680Ω   R3 33kΩ   R4 680Ω
R5 10kΩ   R6 10kΩ   R7 33kΩ

Transistors
TR1, TR2, TR3   All BC108 or BC109

Potentiometer
VR1 10kΩ linear skeleton preset type

Batteries
B1 and B2   Both 6V type PJ996

Fig. 49, page 48

Resistors
R1 4.7kΩ   R2 4.7kΩ R_F 100kΩ

Integrated circuit
IC1 type 741

Battery
18V total (two PP4 in series)

Fig. 50, page 49

Resistors
R2 1MΩ R_F 1MΩ
IC1 type 741

Battery
18V total (two PP4 in series)

Fig. 53b, page 50

Resistors
R1 1MΩ   R2 2.2kΩ   R3 1kΩ

Integrated circuit
IC1 type 741

Diode
D1 type 1N914 or 1N4148

Capacitors
C1   470pF any type   C2 0.1μF polyester

Potentiometer
VR1 2.2MΩ skeleton preset type

Battery
18V (two PP4 in series)

Fig. 54, page 50

Resistors
R1 4.7kΩ   R2 4.7kΩ

Integrated circuit
IC1 type 741

Capacitor
C1 470pF (any type)

Battery
B1 18V total (two PP4 in series)

Fig. 55, page 50

Resistor
R1 2.2kΩ

Integrated circuit
IC1 type 741

Capacitors
C1 and trimmer found by experiment
C2 0.05µF any type

Diodes
D1 and D2 both 1N914 or 1N4148

Battery
Switch if required or use a clip lead.   18V (two PP4 in series)

Fig. 59, page 51

Resistors
R1 2.2kΩ   R2 2.2kΩ   R3 4.7kΩ

Integrated circuit
IC1 type 741

Capacitors found from formulae given

Battery
18V (two PP4 in series)

Fig. 60a, page 52

Resistors
R1 2.2kΩ   R2 2.2kΩ   R3 4.7kΩ

Integrated circuit
IC1 type 747

Capacitors
C1 0.22µF   C2 0.22µF   C3 0.022µF

Battery
18V (two PP4 in series)

Fig. 60b, page 52

Resistors
R1 4.7kΩ   R2 47kΩ

Integrated circuit
IC1 type 741

Capacitors
C1 0.1μF   C2 0.1μF (any type)

Battery
18V (two PP4 in series)

Fig. 60c, page 52

Resistors
R1 4.7kΩ   R2 4.7kΩ

Integrated circuit
IC1 type 741

Capacitors
C1 0.01μF   C2 0.0015μF (any type)

Battery
18V (two PP4 in series)

Fig. 61, page 52

Resistors
R1 4.7kΩ   R2 4.7kΩ   R3 4.7kΩ
R4 4.7kΩ   R5 47 kΩ   R6 47 kΩ

Integrated circuit
IC1 type 741

Capacitors
C1 0.1μF   C2 0.01μF   C3 0.1μF
C4 0.01μF   C5 47μF   (electrolytic)

Potentiometers
VR1 100kΩ linear spindle control
VR2 100kΩ linear spindle control

Battery
18V (two PP4 in series)

Fig. 62, page 53

Resistors
R1, R2, R3, R4, R5   All 100kΩ
R6, R7   Both 47kΩ

Integrated circuit
IC1   type 741

Capacitors
C1, C2, C3, C4   All 4.7μF electrolytic
C5, C6   Both 47μF electrolytic (any type)

Potentiometers
VR1, VR2, VR3, VR4   All 47kΩ or 50kΩ log types
VR5   1MΩ log type   (all spindle controls)

Battery
12V (two PP1 in series)

Fig. 63a, page 53 (add to selected 741 circuit)

Resistors
R1 33kΩ   R2 33kΩ   R3 1kΩ   R4 1.5kΩ

Transistor
TR1 BC109

Battery
B1 12V

Fig. 63b, page 53 (add to selected 741 circuit)

R1 1kΩ   R2 2.2kΩ   R3 680Ω

Transistor
TR1 BC109

Battery
12V

Fig. 65, page 55

Resistor
R1 9.1kΩ (or two 18kΩ in parallel)

Integrated circuit
IC1 NE555 or equivalent

Capacitors
C1 0.01μF   C2 0.01μF   (any type)

Battery
B1 6V type PJ996

Fig. 78, page 67

Resistors
R1 47 or 50Ω   R2 220 or 240Ω   R3 120Ω

Capacitors
C1 2,200μF 25V   C2 1,000μF 25V
C3, C4   Both 470 or 500μF 10V

Diodes
D1, D2, D3, D4   All 1N4001

Transformer
T1 Mains voltage primary, 6 $-$ 0 $-$ 6V 1A secondary

Fig. 79, page 69

Resistor
R1 39Ω 1 watt

Capacitors
C1 470 or 500 μF 10V
C2 220μF 10V

Diodes
D1, D2, D3, D4   All 1N4001
Zener diode 6.2V 1.3W

Transformer
T1 Mains voltage primary, 9V 0.5A secondary

Fig. 79b, page 69

Resistors
R1 47Ω 2 watts   R2 15Ω 1 watt

Capacitors
C1 470 or 500µF 10V
C2 470 or 500µF 10V
C3 0.22µF any polyester type
C4 220 or 250µF 10V

Diodes
D1, D2, D3, D4   All 1N4001
D5   Zener 6.2V 1.3W
D6   Zener 5.1V 400mW

Transformer
T1   Mains voltage primary, 9V 0.5A secondary

Fig. 80, page 70

Resistors
R1 680Ω   R2 1kΩ$^+$ or 1.5kΩ*

Capacitors
C1 1,000µF 25V   C2 220 or 250µF 10V
C3 220 or 250µF 10V   C4 0.022µF polyester type

Diodes
D1, D2, D3, D4   All 1N4001
D5 Zener 6.2V$^+$ or 9.1V* 400mW

Transistor
TR1 AD161 and suitable heat sink or aluminium plate

Transformer
T1 Mains voltage primary, 6V$^+$ or 9V* secondary 0.5A or 1 Amp

($^+$) for 6V output;   (*) for 9V output

Fig. 81, page 71

Resistors
R1 100Ω   R2 2.2kΩ   R3 1.5kΩ   R4 2.2kΩ
R5 1.2kΩ   R6 18Ω   R7 1.5kΩ   R8 1.5kΩ   R9 2.2kΩ

Capacitors
C1 1,000µF 25V
C2 220 or 250µF 10V   C3 0.22µF polyester

Diodes
D1, D2, Both 1N4001
D3 Zener 9.1V 1.3W

Transistors
TR1 AD162 and heat sink or aluminium plate
TR2, TR3, Both BC109
TR4 AD161 and heat sink or aluminium plate

Potentiometer
1kΩ skeleton preset type

Add suitable 12V mains transformer and four 1N4001 diodes

Fig. 82, page 71

Resistors
R1 100Ω   R2 1.5kΩ   R3 18Ω 2 watts
R4 330Ω   R5 2.2kΩ   R6 6.8kΩ

Diode   Transistors
Zener type 6.2V 1.3W   TR1 AD161 and heat sink
                               TR2 BC109

Capacitors
C1 1,000µF 25V
C2 220µF   25V   C3 0.1µF any type
C4 220µF   25V   C5 0.22µF any type

Potentiometer
1kΩ skeleton preset type

Add suitable 12V mains transformer and four 1N4001 diodes

Fig. 83a, page 72

Resistor
R1 22Ω 2 watts

Diodes
D1, D2   Both 1N4001
D3, D4   Both Zener 6.2V 1.3W

Capacitors
C1 1,000µF 25V
C2, C3   Both 220µF 10V
C4, C5   Both 0.1µF polyester

Transformer
T1   Mains voltage primary, 12-0-12V 0.5 or 1A secondary

Fig. 83b, page 72

Resistors
R1, R2   Both 680Ω
R3, R4   Both 2.2kΩ

Diodes
D1, D2, D3, D4   All 1N4001
D5, D6   Both Zener 9.1V 400mW

Transistors
TR1, TR2   Both AD161 and heat sink or aluminium plate

Capacitors
C1, C2   Both 1,000µF 25V
C3, C4   Both 220µF 10V
C5, C6   Both 220µF 10V
C7, C8   Both 0.1µF polyester

Transformer
T1   Mains voltage primary, 12-0-12V 0.5 or 1A secondary

Fig. 86a, page 77

Resistors
R1 1kΩ   R2 68kΩ   R3 680kΩ
R4 220kΩ   R5 18kΩ   R6 1.5kΩ

Transistors
TR1, TR2, TR3   All BC109 or BC108

Battery
6V type PJ996

Fig 86b, page 77

Resistors
R1 4.7kΩ   R2 33kΩ   R3 10kΩ
R4 2.2kΩ   R5 330Ω   R6 680kΩ

Transistors
TR1, TR2   Both BC109

Capacitors
C1 4.7µF   C2 25µF   C3 50µF

Battery
6V type PJ996

The circuits shown in Figs. 90 to 96 call for critically designed printed circuit boards, described in the text.

Characteristics of tuning coils and tuned circuit capacitors depend on requirements; see specific published designs of similar projects. The following lists are for experimental purposes only.

Fig. 90, page 83

Resistors
R1 150Ω   R2 330Ω   R3 33kΩ   R4 1.5kΩ

Integrated circuit
IC1 type TBA651

Capacitors
C1, C2   Both 0.001µF   C3 22µF electrolytic 25V
C4, C5   Both 0.001µF   C6 0.1µF
C7, C8   Both 0.1µF   C9 10µF electrolytic 25V
C9, C10 Both 1nF   C11 0.1µF
C12 100µF electrolytic 25V
C13 22µF electrolytic 25V
C14 1µF
C15, C16 0.0047µF ceramic

Potentiometer
VR1 10kΩ log type

Add tuned circuit components

Fig. 91, page 84

Resistors
R1 100kΩ   R2 56Ω   R3 150Ω   R4 2.2Ω 5W

Integrated circuit
IC1 TBA800 and heat sink

Capacitors
C1 0.1µF   C2 100µF 10V   C3 100µF 25V
C4 1,000pF polyester   C5 100µF 25V
C6 0.05µF polyester   C7 470 or 500µF 25V
C8 0.1µF polyester   C9 100µF 25V

Potentiometer
VR1 10kΩ log type spindle control

Loudspeaker   Battery
LS1 15 ohms   12V (two PP1 in series)

Fig. 93, page 85

Resistors
R1 22Ω   R2 100Ω 1W   R3 2.2Ω 5W

Integrated circuit
IC1 TBA810

Capacitors
C1 0.1µF   C2 470 or 500µF 10V
C3 100µF 25V   C4 4,700pF polyester
C5 820pF polystyrene   C6 100µF 25V
C7 0.05µF polyester   C8 0.1µF polyester
C9 100µF 25V

Potentiometer
VR1 100kΩ log type spindle control

Loudspeaker   Battery
LS1 4 ohms   12V (two PP1 in series)
                        or 18V (two PP4 in series)

Fig. 94, page 87

Resistors
R1 1kΩ   R2 2.2Ω 1W   R3 100kΩ

Integrated circuit
IC1 TDA2020

Capacitors
C1 4.7μF electrolytic   C2 10μF electrolytic
C3 56pF polystyrene   C4 0.047μF polyester
C5, C6 Both 0.1μF polyester

Potentiometer
VR1 100kΩ log type spindle control

Loudspeaker
LS1 4 ohms

Battery
18V (two PP9 in series)

Fig. 95, page 87

Resistors
R1 390Ω   R2 22kΩ   R3 2.2kΩ   R4 470kΩ
R5 150Ω   R6 10kΩ   R7 10kΩ

Integrated circuit
IC1 CA3090

Capacitors
C1 0.0039μF C2 1μF   C3 0.47μF   C4 0.22μF
C5 0.001μF   C6 25μF C7 1μF   C8 6,800pF   C9 6,800pF

Plus 2mH tuned inductor and pilot lamp

Battery
18V (two PP4 in series)

Fig. 96, page 88

Resistors
R1 10kΩ   R2 33kΩ   R3 3.9kΩ
R4 470Ω   R5 100kΩ

Integrated circuit
IC1 CA3089

Capacitors
C1 0.022µF   C2 0.022µF
C3 100pF ceramic   C4 0.33µF polyester

Plus tuning meter (150µA) and tuned inductors or ceramic filters.

Potentiometer
VR1 470kΩ or 0.5MΩ log type

Battery
12V (two PP1 in series)

AMATEUR RADIO
G3VFV

CHAS H. YOUNG LTD.,
170 Corporation Street, Birmingham, B4 6UD
021 - 236 - 1635

A wide selection of electronic components available from stock—resistors, capacitors, coils, transistors, relays, TRIACS, SCRs, potentiometers, transformers, soldering irons, etc.

AGENTS FOR Eddystone Communication Receivers, Microwave Modules, Icom Equipment and Finnigans Hammerite Paints.

Mail Order Service   Access & Barclaycard

Questions and Answers

Each of these books contain simple and concise answers to questions which puzzle the beginner and the student — from first principles to a useful level of practical knowledge. Why not write for further information today?

QUESTIONS AND ANSWERS ON COLOUR TELEVISION 2nd edn
J.A. Reddihough and D. Knight

A simple practical account of colour television transmission and reception for the enthusiast, technician and service engineer. Covers the techniques used in PAL receivers and includes the principles of colour; transmission and reception of compatible colour signals; the shadowmask, Trinitron and PI tubes and modulation circuits; convergence circuits, convergence adjustments.

134 pages *illustrated* 0 408 00162 3

QUESTIONS AND ANSWERS ON ELECTRONICS
Clement Brown

A condensed account of a wide-ranging subject, intended to give the interested layman and the student an insight into the underlying principles and numerous applications of electronics.

114 pages *illustrated* 0 408 00041 4

QUESTIONS AND ANSWERS ON HI-FI
Clement Brown

A practical survey of the field of high fidelity sound reproduction intended to guide amateur enthusiasts interested in serious listening and to serve more advanced students as a reminder of modern practice, the emphasis being on engineering aspects. The notes on principles underlying recent developments will convince readers that it is worth while going to a certain amount of trouble to obtain better results. The book covers all hi-fi sources — disc, tape and radio, and guidance is given on costs, specifications and planning requirements.

104 pages *illustrated* 0 408 00151 8

QUESTIONS AND ANSWERS ON INTEGRATED CIRCUITS
R.G. Hibberd

Covers all the main types of integrated circuits — thick and thin film, monolithic and hybrid, digital and linear — and also deals with Boolean algebra and binary notation. Resistor, diode and transistor logic circuits are described and compared and typical applications are discussed.

96 pages *illustrated* 0 408 00115 1

QUESTIONS AND ANSWERS ON RADIO AND TELEVISION 4th edn
H.W. Hellyer and I.R. Sinclair

This new edition has been extensively revised and includes new illustrations and a new section on the use of transistors.

128 pages *illustrated* 0 408 00249 2

QUESTIONS AND ANSWERS ON TRANSISTORS 3rd edn
Clement Brown

Starting with the properties of semiconductor materials, this book provides a readable account of the operation and uses of transistors and other associated devices. A section is included on servicing transistorised equipment, with advice on the precautions necessary in dealing with transistors.

96 pages *illustrated* 0 408 08161 9

 Newnes Technical Books
Butterworths, Borough Green, Sevenoaks, Kent TN15 8PH

Index

A.M. radio, 82
Amplifier, 43, 74, 80
Analogue converter, 43, 54
Astable multivibrator, 16, 27, 31, 56

Battery, 62
Binary notation, 19
Bias, 75, 77
Bistable switch, 12, 28, 34, 54
Buffer amplifier, 24, 54

Changeover switch, 12
Clock pulse generator, 18, 61
Collector current, 9
Comparator, 45, 54
Cross-coupling, 12, 28, 37
Current gain, 9

Darlington pair, 45
Differential amplifier, 45, 70
Diode-transistor logic (DTL), 21, 40, 68
Direct coupling, 77
Dual in-line package (DIL), 40

Fan out, 23
Feedback, 47, 77, 79, 82
Filter, 66
Flip-flop, 23
F.M. radio i.c., 88

Gates, 20, 30, 61
Ground plane, 80

Heatsink, 10, 80, 85

Integrated circuits, 21, 37, 46, 68, 80
Interaction, 43

Light emitting diode, 25
Linear circuits, 43
Link wires, 37, 80
Logic code, 19
Logic indicator, 25
Logic switch, 15
Long-tailed pair, 45

Metronome, 18
Missing pulse detector, 58
Modules, 30
Monostable multivibrator, 28, 34, 60
Multivibrator, astable, 16, 27, 31, 56

Negative logic, 19

On-off switch, 12
Operational amplifier, 43, 68, 70, 81

Positive logic, 19
Power amplifier, 85
Power supplies, 62
Printed circuits, 41, 79
Pulse position modulator, 59
Pulse width modulator, 58

Quad in-line package, 82

Rectifier, 64
Resistor-transistor logic (RTL), 22

111

Schmitt trigger, 37
Short-circuit protection, 69
Smoothing, 64
Speed-up, 31
Square wave generator, 16
Stabiliser, 68
Staircase generator, 34
Stereo decoder, 87
Switching circuits, 2, 8

Television, 74
Thermal stability, 74

Threshold switch, 44, 54
Timer, 54
Transformer, 63
Transistor switch, 2, 8
Transistor-transistor logic (TTL), 22, 68
Tuned circuits, 84

Wheatstone bridge, 44
Wiring layouts, 26, 35

Zener diode, 68

The easy way to a PCB...
...the Seno 33 system!

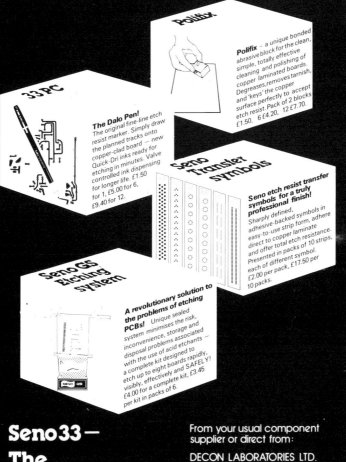

Polifix — a unique bonded abrasive block for the clean, simple, totally effective cleaning and polishing of copper laminated boards. Degreases, removes tarnish, and 'keys' the copper surface perfectly to accept etch resist. Pack of 2 blocks £1.50, 6 £4.20, 12 £7.70.

The Dalo Pen! The original fine-line etch resist marker. Simply draw the planned tracks onto copper-clad board — new Quick-Dri inks ready for etching in minutes. Valve controlled ink dispensing for longer life. £1.50 for 1, £5.00 for 6, £9.40 for 12.

Seno etch resist transfer symbols for a truly professional finish! Sharply defined, adhesive-backed symbols in easy-to-use strip form, adhere direct to copper laminate and offer total etch resistance. Presented in packs of 10 strips, each of different symbol. £2.00 per pack, £17.50 per 10 packs.

A revolutionary solution to the problems of etching PCBs! Unique sealed system minimises the risk, inconvenience, storage and disposal problems associated with the use of acid etchants — a complete kit designed to etch up to eight boards rapidly, visibly, effectively and SAFELY! £4.00 for a complete kit, £3.45 per kit in packs of 6.

Seno 33 —
The Laboratory in a box

From your usual component supplier or direct from:

DECON LABORATORIES LTD.
Ellen Street, Portslade,
Brighton BN4 1EQ
Telephone: (0273) 414371
Telex: IDACON BRIGHTON 87443

All prices post & VAT inclusive. Data sheets free of charge.

Still the *simplest, fastest* way of constructing your electronic circuits. Vero Electronics Ltd manufacture boards to suit most of your requirements, including, Verostrip, Dual in line I.C. boards, Plain Boards, etc. A new catalogue is available (price 10p + S.A.E. at least 7" x 9")

vero
VEROBOARD®

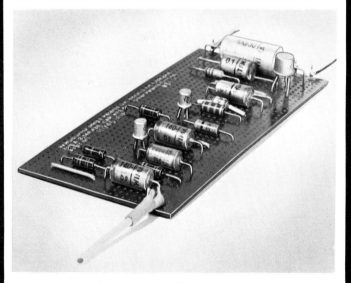

**Vero Electronics Limited, Retail Dept.,
Industrial Estate, Chandler's Ford, Hants., SO5 3ZR
Telephone: Chandler's Ford 2956 (STD 04215)**

TOOLS

AUDIO KITS

IC's

BLANK CASSETTES

TRANSFORMERS

DORAM

JUST ABOUT EVERYTHING FOR THE AMATEUR RADIO, ELECTRONICS AND HI-FI ENTHUSIAST

All these goodies plus many more interesting projects may be found in the Doram catalogue - only 60p

DIN PLUGS & SOCKETS

DORAM ELECTRONICS LTD, P.O. BOX TR8, LEEDS, LS12 2UF

CASES

SPEAKERS

SWITCHES

MICROPHONES (all types)

Get a great deal from Marshall's

A Marshall (London) Ltd
40/42 Cricklewood Broadway London NW2 3ET Tel: 01-452 0161/2 Telex: 21492
& 85 West Regent St Glasgow G2 2QD Tel: 041-332 4133
& 1 Straits Parade Fishponds Bristol BS16 2LX Tel: 0272-654201/2
& 27 Rue Danton Issy Les Moulineaux Paris 92
Call in and see us 9.-5.30 Mon-Fri 9.-5.00 Sat
Trade and export enquiries welcomed
Catalogue price 55p (40p to personal callers)

LONDON • BRISTOL •GLASGOW•

Top quality products!

We carry a larger and more interesting range than most other UK distributors and stock 8,000 lines of products all approved by leading manufacturers and guaranteed to their specifications — Siemens, SSDI, Sescocem, ITT, Motorola, Mullard, Plessey, National, Piher, etc.

Top quality service!

Where else can you obtain all you need from one source? We invite all enquiries and welcome personal callers to any of our locations. Call in and collect your order immediately.

An informative catalogue!

With 158 pages crammed with technical information, pin connections and many interesting new products. Only 55p post paid or 40p to personal callers.

Competitive prices!

Just compare ours with other dealers. Where else can you get a deal as good as this backed up by our excellent service?

All you need!

All from the same place.

BI-PAK Semiconductors

OFFER AN EXTENSIVE RANGE OF ELECTRONIC COMPONENTS AND EQUIPMENT AT THE MOST COMPETITIVE PRICES...

Including
- TRANSISTORS
- DIODES
- S.C.R.'s
- TRIACS
- I.C'S
- I.C. SOCKETS
- RESISTORS
- CAPACITORS
- VEROBOARDS
- P.C. BOARDS
- P.C. PENS ETC.
- INSTRUMENT CASES
- ALUMINIUM BOXES

- SOLDERING IRONS
- SOLDER
- CABLE
- PLUGS & SOCKETS
- AUDIO LEADS
- AUDIO ACCESSORIES
- POTENTIOMETERS
- SWITCHES
- L.E.D. DISPLAYS
- TRANSFORMERS
- TECHNICAL BOOKS
- VOLTAGE REGULATORS
- CASSETTES

Plus an Extensive Range of Audio Modules

Send 50p for our Comprehensive Catalogue
To
**BI-PAK SEMICONDUCTORS
P.O. BOX 6
WARE
HERTFORDSHIRE SG12 9AD**

LOOK NO FURTHER!

FOR YOUR COMPONENT NEEDS FOR PROJECTS IN THIS BOOK...

PHONE

Semiconductor Supplies (Croydon) Ltd.

SEND S.A.E. FOR FREE CATALOGUE

Orchard Works, Church Lane, Wallington, Surrey SM6 7NF

Tel. 01-647 1006 (5 lines)
Telex: 946650

AND OUR ASSOCIATE COMPANY

AMATEUR COMPONENTS

ORCHARD WORKS, CHURCH LANE, WALLINGTON, SURREY SM6 7NF

MAIL ORDER DIVISION OF SEMICONDUCTOR SUPPLIES (CROYDON) LTD.

For Semiconductors Capacitors Resistors I/C Sockets L.E.D.s and Hi-Fi Accessories

BULK ORDER DISCOUNT

Beginner's Guides

These books are intended to provide a basic understanding of a wide range of subjects. Some related titles appear below. If you would like to know more about these and other titles in the series the publishers will be pleased to give further information.

BEGINNER'S GUIDE TO AUDIO
Ian R. Sinclair
Covers the working principles of devices ranging from the microphone to the loudspeaker and examples of the electronic circuits which are used.

192 pages   *illustrated*   *0 408 00274 3*

BEGINNER'S GUIDE TO COLOUR TELEVISION 2nd edn
Gordon J. King
The colour television system adopted in Great Britain, parts of Europe and some countries in other parts of the world is called PAL (phase alternate, line). This system is a refinement of that used in the USA and some other countries, known as NTSC. The reader is guided through the principles of NTSC and PAL to an understanding of the method of operation of the PAL system from aerial to display tube.

198 pages   *illustrated*   *0 408 00101 1*

BEGINNER'S GUIDE TO INTEGRATED CIRCUITS
I.R. Sinclair
This book is for the comparative newcomer to electronics, with some knowledge of transistor circuits, wishing to acquire an understanding of i.c.s. The principles and construction of i.c.s are described as well as many examples of practical i.c. circuits.

192 pages   *illustrated*   *0 408 00278 6*

BEGINNER'S GUIDE TO RADIO 8th edn
Gordon J. King
Introduces the reader in easy step-by-step stages to all aspects of radio technology, from simple electromagnetic theory to the full range of radio components and circuits. Completely rewritten and updated, the 8th edition contains all the latest developments in radio technology.

240 pages   *illustrated*   *0 408 00275 1*

BEGINNER'S GUIDE TO TELEVISION 5th edn
Gordon J. King
This edition records the most up-to-date developments in the world of television, including useful information on the transistor circuits being used in contemporary receivers and the new breed of single standard receivers.

212 pages   *illustrated*   *0 408 00349 9*

BEGINNER'S GUIDE TO TRANSISTORS 2nd edn
J.A. Reddihough. Revised by **I.R. Sinclair**
Provides an excellent and thorough introduction to the transistor in an easily understood form that will be welcomed by all who are new to the field. It shows how the transistor is used in everyday practical circuits and introduces the various techniques involved. The approach throughout is non-mathematical and the emphasis is on circuit operation rather than design.

162 pages   *illustrated*   *0 408 00145 3*

 Newnes Technical Books
Butterworths, Borough Green, Sevenoaks, Kent TN15 8PH

Practical Electronics is the magazine for all home constructors. Every month there is a wide range of useful constructional projects which bring the latest technology within reach of the electronics hobbyist. There are also general features as well as news and comment.

Progress with
PRACTICAL ELECTRONICS
every month 45p.